JANUA LINGUARUM

STUDIA MEMORIAE
NICOLAI VAN WIJK DEDICATA

edenda curat

C. H. VAN SCHOONEVELD
Indiana University

Series Minor, 129

LINGUISTIC STRUCTURES IN SCIENTIFIC TEXTS

by

MYRNA GOPNIK
McGill University

1972
MOUTON
THE HAGUE · PARIS

LIBRARY OF CONGRESS CATALOG CARD NUMBER: 72-190147

Printed in Belgium by NICI, Printers, Ghent.

For Irwin and for Alison, Adam, Morgan, Hilary, Blake and Melissa: for advice and encouragement but more important for reality and fantasy; for truth and games.

TABLE OF CONTENTS

1. *Introduction* 9
  1.0 Selection of texts 10
    1.1 Stylistic uniformity 11
    1.2 Explicit semantic content 12
    1.3 Texts 12
  2.0 The concept of text 12
  3.0 Paraphrase 15
    3.1 Included paraphrase 15
    3.2 Normalization 16

2. *Decomposition and normalization* 18
  1.0 Introduction 18
    1.1 Intrasentential structures 18
    1.2 Intersentential structures 19
      1.21 Pro-word interreference 20
      1.22 Zeroings and deletions 20
      1.23 Resolution of ambiguities 21
  2.0 Texts 22
  3.0 Pro-words 23
    3.1 *We* and passives 23
    3.2 *It* 25
    3.3 Ordinal pronouns 25
      3.31 Demonstratives 26
      3.32 *The* 28
  4.0 Patterns of word reoccurrence 30
  5.0 Decomposition 31

5.1 Coordinating conjunctions 31
5.11 *And* 31
5.111 Standard *and* 31
5.112 Temporal *and* 33
5.113 Compound subject *and* 35
5.114 *And* from rearrangement 36
5.12 *Or* 37
5.121 Exclusive vs. inclusive 37
5.122 Truth-functional criteria 37
5.123 Structural criteria 39
6.0 Passives 40
7.0 Comparative structures 42
7.1 Structure of noun phrases 43
8.0 Summary 44

3. *Syntactic structures in texts* 46
1.0 Definition of criteria for structure 46
1.1 Occurrence of transformations 48
1.2 Position in text 49
1.3 Non-occurrence of transformations 49
1.4 Interrelationships among transformations . . . 50
1.5 Interrelationships among noun phrases and verb phrases 51
2.0 Major structures 52
2.1 Controlled experiment 53
2.2 Hypothesis verification 53
2.3 Technique-descriptive 54
3.0 Controlled experiment 54
3.1 Establishment of classes 57
3.2 Parenthetical structures 58
3.3 Measurement statement 60
3.4 Comparative structures 63
3.5 Comparatives and classes of variables 71
3.6 Container structures 72
3.7 Summary of controlled experiment structure . . 74
4.0 Hypothesis verification 77

4.1 Single hypothesis texts 78
4.2 Hypothesis statement 80
4.21 Citation-container 80
4.22 Law-like sentences 82
4.23 Alternate hypotheses 84
4.3 Intention sentences. 85
4.4 Conclusion sentences 86
4.5 Evidence structures 88
4.6 Example analysis 89
5.0 Technique-descriptive 92
5.1 Temporal structures 92
5.11 Adverbs of time 92
5.12 Successively modified noun phrases . . . 94
6.0 Aberrant texts 96

4. *Paraphrastic analysis of texts* 98
1.0 Possible techniques 99
1.1 Extraction and recombination 99
1.2 Structural patterning 101
1.3 Frequencies 105
2.0 Choice among paraphrases 107
2.1 Storage and retrieval 107
2.2 Summary — abstract 108

Appendix I — Full texts used in study 110
Appendix II — Patterns of pro-word interreference in 16 texts 127
Appendix III — Pattern of occurrence of words (except for grammatical constants) which occur two or more times in original text 133
Appendix IV — Decomposition of texts 135
— Removal of binaries

Bibliography 146

1

INTRODUCTION

This study will investigate certain formal properties of short discourses in order to establish for these discourses a computable underlying form. It is therefore an extension of current work in linguistic theory which attempts to establish the underlying form for single sentences by a series of explicit rules. As an extension of this theory this study will refer to those rules developed for the analysis of single sentences, but since the rules for the analysis of single sentences are themselves at present still being developed it is to be expected that their present form will be amended as the result of further investigation. Therefore, we have deliberately tried to indicate the general patterns of certain transformations rather than the precise details of all the rules involved. This direction of proceeding is warranted because the final goal of this study is rather to provide a theoretical framework in which problems of the analysis of texts can be formulated than to establish a particular set of rules to be used in this analysis. From this point of view the important outcome of this study is the generalized pattern of investigation of the structure of texts, though of course, the empirical results of this work are in themselves useful indicators of the kinds of results such studies can produce.

In order to specify the types of intersentential connections which hold between the sentences in a text it is necessary to assume that not every sequence of sentences produces a coherent text. It is clear that without this assumption no analysis would be possible, or rather the results of such an analysis would be trivial in that the only rule for the structure of texts would be that each text be segmentable into a sequence of sentences. It would seem clear that the assumption that some sequences of sentences are texts while other sequences of sentences are not texts is in accord with our intuitive knowledge about the characteristics of language. Further any adequate theory of linguistics must provide some way of characterizing the sequences of sentences which are texts as opposed to the sequences of sentences which are not texts. Moreover it is clearly desirable that other generalizations which hold about texts be likewise characterizable.

One of the important generalizations to be characterized is the fact that some sets of texts are semantically equivalent. Certainly not all cases of semantic equivalence of texts need be characterized. Some of these cases of semantic equivalence may be the result of extra-linguistic facts and therefore not amenable to linguistic description, but other cases of semantic equivalence may be the result of certain linguistic operations and therefore should be describable in linguistic terms.

Underlying both of these assumptions is the further assumption that texthood and text paraphrase can be described in terms of formal properties which are representable in a set of rules. That is, that the intuitive judgements about texthood or text paraphrase are, at least in part, the result of the application of certain rules of analyses which are discoverable and specifiable in terms of explicit rules.

1.0 SELECTION OF TEXTS

This study in order to investigate the structure of short discourses will use articles selected at random from the Proceedings of the Federation of American Societies for Experimental Biology (FASEB) as its basic data. Each text is approximately 250 words long and is presented as a complete text. The FASEB proceedings was selected as a source for the texts for certain theoretical and practical considerations. One of the primary requirements for this analysis was that the texts concerned be as free as possible from stylistic variation. A second strong requirement was that the semantic content of the texts be as explicitly stated as possible and as free from outside references as possible. A third consideration was that the texts be long enough to show some structural patterns, but short enough to be easily analyzed.

1.1 *Stylistic uniformity*

The requirement that the texts be stylistically simple and uniform will make transformational analysis of the text both simpler and more generally useful. If the texts are stylistically simple then the set of transformations which will be needed to perform the analysis will be well entrenched in the language and the analysis will therefore be able to be applied to a broad range of documents. By expanding the set of transformations used, this analysis should be applicable to a broader range of documents. A basic assumption of this analysis, or indeed of any analysis which makes use of some sort of normalization of the text is that the original text is equivalent in the respect appropriate to the analysis (e.g., semantically equivalent), to the normalized text and, therefore, that an analysis of the normalized text is an appropriate substitute for an analysis of the original text. That is, what is interesting about this text is not the particular set of transformations which generated this text, but the information which is carried by these transformations. Therefore any transformation which is information preserving, i.e. paraphrastic, performed on any sentence in the text will produce a new text which is equivalent to the original text in its information content. It is assumed that the texts in the FASEB proceedings are essentially reports of information and are therefore appropriately handled by a normalization procedure.[1] A corollary of this assumption is that, therefore, the style of the texts will be free

[1] This assumption would not hold true for literary texts in which the particular linguistic structure of the text is at least as important as its semantic content. For a brief discussion of some of the problems of paraphrase in literary texts, see "Toward a General Theory of Style: The Law of Maximal Form", with Irwin Gopnik, *Style*, Vol. 1, Fall 1967, No. 3.

from any non-normal transformations which could in any way obscure the content of the texts. Moreover, since the texts are all reports in the same field a certain uniformity of style is to be expected.

1.2 *Explicit semantic content*

The second requirement which calls for the semantic content of the texts to be as explicit as possible appears to be best satisfied by very technical texts. The range of meanings and the set of synonyms for the terms in these texts are relatively well defined. In this sense it is certainly easier to determine the meaning of a technical word like *cyclophosphoramide* than of a non-technical word like *justice*. Since technical terms have this well defined range of meanings as opposed to the broader range of meanings of non-technical terms the problems of reference and the choice between possible ambiguous transformational routes is somewhat reduced.

1.3 *Texts*

The texts chosen are all approximately 250 words long. They are mostly one paragraph texts though there are a few two paragraph texts in the sample. It is important to note that though they are short they are complete. They are neither excerpts from longer texts nor abstracts from longer texts. If they were abstracts of a longer text then one might expect that the relationships among the sentences might reflect the relationship between the longer stretches of text from which they were abstracted. If they were excerpts from some longer text then one might expect that the pattern of interrelationships among the portion of text under consideration could be determined by the structure of the longer text.

2.0 THE CONCEPT OF TEXT

One of the basic concepts of this study is the notion of TEXT. Since text is an assumed primitive concept of this study we will not attempt to give any rigorous definition of text. We will, however, present a general discussion of the basic concepts related to the notion of text which may serve to elucidate the range of meaning of the term. There are certain organized sets of sentences which have a certain independence from any other sentences. One of the conditions for this independence seems to be that almost all of the questions of reference are resolved within the text. (There are, for instance, certain anaphoric *the*'s which may not be resolved within a text and yet the set of sentences is still a text. These will be discussed in detail in Chapter 2.) Though the resolution of reference may be a necessary condition for texthood, it is not a sufficient condition. Not any set of sentences which do not have any unresolved references can be said to form a text.

There are two levels of criteria upon which texthood could be judged, syntactic and semantic. In the first place, there are syntactical limitations on the sentence structures of the text. That is not to say that not all structures can occur in a text, but only that a text may contain syntactic dependencies. A certain structure of a sentence at one point in the text may demand a certain structure at some other point in the text. Moreover, there may be some restrictions on the order of syntactic structures within a text. An example of such syntactic restrictions is, for instance, that a sentence of the structure

The former V_1 and the latter V_2.

cannot occur as the first sentence of a text. As we will see later in this study there is more than one pattern of interrelationships among the syntactic structures of the sentences in a text which can produce an acceptable text. A major portion of this study will concern itself with the discovery and description of these syntactic interrelationships.

In addition to these syntactic requirements for texthood there are also some semantic requirements. The concepts in a text must be "talking about the same thing". That is to say, there must be a certain semantic cohesiveness among the sentences of a text. In some cases this semantic cohesiveness can be determined by the syntactic forms of the sentences involved. One can, for instance, establish this cohesiveness on the reoccurrence of noun phrases and verb phrases throughout the text. If the same noun phrase occurs as the subject of every sentence in the text then this pattern of reoccurrence of identical segments in identical syntactic roles can be used to establish this cohesiveness. In some cases this reoccurrence may not occur explicitly in the text itself, but is implicit in the pattern of reference and in the zeroed portions of the text. In many cases, however, there is no pattern of reoccurrence which can be used to establish the semantic interrelationships of the text. These relationships can be established only by finding out what the text means. As we will see later in Chapter 3 and Chapter 4 there are some cases in which the structure of the text itself establishes certain semantic interconnections among some of the terms in the text. In general, this study will not concern itself with the problem of the strong semantic analysis of texts except insofar as the analysis can be done in terms of the syntax and structure of the text. It will be concerned mainly with the problems of the weak semantic, that is to say paraphrastic, analysis of the text.

The concept of TEXT used in this study is in many ways analogous to the concept of SENTENCE used in transformational grammars. That is, it is assumed for this study that certain sequences of sentences are definitely texts, and that certain other sequences are definitely non-texts. Rules will be devised which will serve to distinguish between these two sets of sequences of sentences. The rules themselves will then differentiate among borderline cases. It should be stated that from this point of view the empirically given texts which form the foundation of this study are not considered to necessarily be perfect texts. They are only a convenient starting point for any analysis. It may be that certain portions of the empirically given texts violate a rule which, in regard to other sequences of sentences, does serve to distinguish texts from non-texts. In these cases that sequence of sentences which violates the rules for forming texts is not itself a text, just as in transformational grammar an empirically given sequence of morphemes may be a non-sentence if it cannot be generated by the rules of the grammar. This study will provide some of the rules for generating texts.

3.0 PARAPHRASE

In order to define the concept of paraphrase it is necessary to distinguish between WEAK SEMANTICS and STRONG SEMANTICS. A strong semantic analysis of a given linguistic segment determines what the segment "means". A weak semantic analysis of a linguistic segment determines whether or not it "says the same thing" as some other linguistic segment. As has been pointed out "This 'saying the same thing' is just a relation between two sentences [or longer texts] and it does not presuppose something else, the 'thing' said in each sentence."[2] It should be pointed out in reference to the weak semantic analysis of texts, that the determination of whether or not two texts 'say the same thing' cannot be done by finding out whether each sentence of one text 'says the same thing' as a corresponding sentence of the other text. As we have mentioned above the organization of sentences in a text imposes certain syntactic and semantic relationships among the sentences of the text, and therefore, what the text 'says' is not simply the sum of what each individual sentence within the text 'says'.

3.1 *Included paraphrase*

If two linguistic segments are semantically equivalent then they are paraphrases of one another. In some situations semantic equivalence is necessary and therefore it will be important to be able to say that two segments are paraphrases of one another. In other situations, however, the strong requirement of semantic equivalence is not necessary. We may only need the relationship of semantic inclusion. For instance, in discovering rules for

[2] H. Hiz, "The Role of Paraphrase in Grammar". *Monograph Series on Languages and Linguistics*, Number 17. Report of the 15th Annual R.T.M. on Linguistic and Language Studies, ed. C.T.J.M. Stuart (April, 1964).

generating one text from another we may not be concerned with generating a text which 'says the same thing' as the original text, but only with generating a text which 'does not say anything which was not said in the original text'. This second requirement is not so strong a requirement as the first since it involves only inclusion in one direction while the first requirement involves inclusion in both directions. This new relationship will be called INCLUDED PARAPHRASE. One linguistic segment is an INCLUDED PARAPHRASE of another segment if it 'says only things that are said in the second segment'. It may not say everything said in the other segment, but it does not say anything new. Paraphrase can be defined in terms of included paraphrase. Two linguistic segments are paraphrases of one another if and only if the first is an included paraphrase of the second and the second is an included paraphrase of the first. That is, semantic equivalance is equivalent to semantic inclusion in both directions.

3.2 *Normalization*

The rules for normalization of a text which will be discussed in this study are designed to create a standard form for the whole set of texts which are paraphrases of one another. Since all of these texts 'say the same thing' the text which has the normal form can be used for any semantic analysis of any member of the paraphrastic set. The differences between these paraphrases arise from certain paraphrastic permutations, substitutions and zeroings. These devices work at both the level of the sentence and at the level of the text. The paraphrastic transformations which work at the level of the sentence are widely discussed in current work in linguistics. However, the devices which work at the level of the text are not as well known and will be specified in detail in this study. In the case of either kind of device the normalized text is derived from the original text by applying certain of these paraphrastic transformations so as to preserve the information content of the text while changing the form of the text to the normal form. By changing the texts to a normal form the underlying patterns of relationship among the various parts of the text can be made more explicit and the subsequent semantic analysis can be made more uniform.

2

DECOMPOSITION AND NORMALIZATION

1.0 INTRODUCTION

In this chapter we will report the results of applying certain transformations to the sentences in the texts under consideration. The chapter itself will present the theoretical background for the treatment of the text and will report on the general patterns found in the texts, using some examples from the texts to illustrate these patterns. The empirical data: texts used, detailed decompositions, structures of comparatives, patterns of word reoccurrence, patterns of interreference, etc. will appear at the end of this study in an appendix so that it can be available for easy reference.

It was decided to use decomposition transformations as part of the analyses of the texts, in the first place, to study certain aspects of transformations in texts and also to provide a normalized text for the subsequent semantic analysis to be conducted on. We will investigate certain patterns of occurrence of transformations which can serve to differentiate among different types of texts. By stating these patterns we may be able to begin to have a set of rules for the analysis and construction of individual sentences into the larger linguistic structure, a text.

1.1 *Intrasentential structures*

In the process of decomposing these texts two different aspects of transformational analyses were elucidated. The first of these involved problems of intrasentential decomposition, the second involved problems of intersentential dependencies in decomposition. Since so much work has already been done in discovering and describing transformations involving sentences in isolation most of the distinctions at this level are reiterations of work which can be found in other sources. In fact, since the other studies of these problems deal with all aspects of one kind of transformation they are usually more exhaustive than is necessary for our purposes. It is, however necessary to make explicit the particular transformations which are used for the decomposition of the texts in question, especially when the transformation used is not the standard one. For instance, the decomposition of sentences in which *and* occurred followed not one, but four different patterns. Some of these different patterns have already been discussed in other papers.[1] There are two compelling reasons for repeating the distinctions found again in this paper. In the first place it is important to make clear the decomposition path taken by each of the sentences in these texts so that the major patterns of transformations in the text as a whole will be clear and in the second place much of the work done on specifying various transformations does not follow a uniform format so clarity of exposition requires making explicit the transformations used at each point in the decomposition process.

1.2 *Intersentential structures*

Though there are many other treatments of the problems involved with intrasentential dependencies, the problems involved in intersentential dependencies have not been dealt with as extensively. These problems involve looking at structures which are more than one sentence long. By studying the structure of texts we will be particularly concerned with these problems. In fact, any analysis of a text is impossible without addressing yourself to precisely these problems. The dependencies between sentences in a text are of two

[1] Irena Bellert, "On Certain Syntactical Properties of the English Connectives *and* and *but*." *Transformations and Discourse Analysis Projects* #64, University of Pennsylvania.

major kinds. In the first kind of intersentential dependency the decomposition of a sentence in a text cannot be done on the basis of that sentence alone, but if the other sentences in the text are considered the decomposition can be done. Examples of this kind of dependency are pro-word reference, zeroings, etc. In the second kind of intersentential dependency the sentences involved can be decomposed independently, but considerations of the structure of one of the sentences has an effect on the decisions made about the structure of the other sentence. An example of this kind of intersentential dependency is the resolution of two possible routes for decomposing one of the sentences by considerations of the structure of the other sentence.

1.21 *Pro-word interreference*

In the case of pro-word reference, which will be discussed in detail later in this chapter, some pro-word constants, e.g. *this, these, it, they*, etc., appears in one of the sentences in the text. Furthermore, this pro-word is not resolvable within the sentence itself, but is resolvable in the larger context of the text. The simplest of these cases is when the pro-word refers to a noun phrase in the previous sentence. The reference, however, can be more complex. The noun phrase referred to may not be the preceding noun phrase, but may actually be a noun phrase two or more sentences away. Almost always such cases involve some semantic considerations. In some cases the semantic information to resolve this reference is contained within the text under consideration, but in some cases the information needed to resolve the problem must come from general knowledge about the world. The phrase referred to is not always a particular noun phrase. The reference may be to a previous whole sentence or to a set of previous sentences, or may even refer to the whole previous text.

1.22 *Zeroings and deletions*

In the case of zeroings and deletions the decomposition cannot be accomplished because some necessary part of the structure of the sentence needed for the decomposition is missing. This structure may not be recoverable from the sentence itself, but, as above, is resolvable in some larger context. One may, for instance, have a sentence *Similar results were observed in vagotomized animals except for the absence of the transient initial increase in rate and minute volume.* (text #197) In order to decompose this sentence we need the structure: *X* is similar to *Y*, but we must go to the previous sentence in the text to establish *Y*.

1.23 *Resolution of ambiguities*

The third type of intersentential dependencies which we will deal with in detail later in this chapter are those cases in which decompositions of each of the sentences can be done individually, but in which a choice of a particular structure of one sentence influences the choice of structure of another sentence. This happens most often when at least one of the sentences has more than one possible decomposition. In these cases the structure of one of the other sentences in the text reinforces one of the decompositions and contradicts the other, and thus resolves the ambiguity in the first sentence. This happens, for instance, in the case of deciding between the exclusive or the inclusive *or*. The structure of one of the following sentences often resolves the ambiguity of the original *or* sentence.

A different kind of intersentential dependency which involves one structure influencing another is the patterns of cooccurrence of structures within any one text. Such cooccurrence restrictions are very hard to substantiate on the basis of empirical data, since the sample may simply not contain an instance of a counter-example, though one may be readily producible. On the other hand, certain general patterns of cooccurrence of types of transformations may be prevalent enough in the sample to allow you to infer that this is an instance of a more general pattern in the language. As we will discuss in more detail later, the evidence of the texts we have studied suggests that such a cooccurrence pattern exists between the occurrence of a container sentence in final position in the text and the occurrence of comparative structures in other parts of the text. If, in fact, such dependencies are prevalent then this would have a significant influence on approaches to the semantic interpretation of these different types of texts.

2.0 TEXTS

At the beginning of this chapter we mentioned that one of the other motivations for performing this decomposition on the texts in question was to provide some normalizations of the texts upon which the semantic analysis could be done. In this way we hope to eliminate some of the purely stylistic variations within each of the texts and, by representing each text as the result of certain transformations, impose a degree of syntactic uniformity on the texts. The decisions about the extent of decomposition which should be done to have this standardized form without sacrificing any structures which are significant for the semantic interpretation of the texts is not either simple or trivial to make. Such a decision involves deciding which structures are merely stylistic variants, and which structures are involved in the determination of the type of the text and, therefore, have semantic significance.

The texts which are analyzed in this study were all selected from brief research reports submitted by scientists who wished to give a presentation at the 1962 meeting of The Federation of American Societies for Experimental Biology. These papers are short summaries (250 words or less) of the intended presentation. Though they are sometimes referred to as "abstracts" these texts are not, in fact, texts which are abstracted from some longer text. They are, in almost all cases, preliminary reports which are later expanded into a longer, more detailed report. For the sample under consideration in this study twenty-eight texts were selected at random from this larger corpus. The assumption we make is that this random selection will be representative of the whole corpus, and that patterns which occur in this smaller corpus will be indicative of the patterns in the larger corpus. There is, however, no intention of equating the patterns of structure of texts in our empirical sample with the extent of possible structures of texts. All we intend to use the empirical sample for is to discover the underlying recursive rules which underlie all texts. From this point of view, obviously, any rules which do not occur within our sample will not be elucidated, but certain of the rules which can be substantiated on this real sample can serve as a foundation for further investigation.

3.0 PRO-WORDS

One of the first analyses that we made on our sample texts was an analysis of the patterns of pro-word interreference. This analysis was a necessary prerequisite before any decomposition of the individual sentences could proceed. And, too, we were interested in finding out how difficult the problem of pro-word reference in scientific texts would be. An outline of the pro-word interreference pattern for the first fifteen texts is given in Appendix II. In all of the texts which we looked at in detail, only two pronouns occurred. They were *we* and *it*.

3.1 "*We*" *and passives*

One of these, *we* was in fact not resolved in the body of the text. It was clear that the referent was to the author of the paper, and if the title and author of the text are considered as part of the text then this referent causes no problems. However, it may be that there is no necessity for clarification of this reference in an analysis of a scientific text. In all of the texts in which *we* did not occur the passive transformation did occur. In all of the instances of the passive transformation the second noun of the structure was deleted. For instance, we find:

Fertile eggs were removed from the incubator after 3 days of incubation. (#018)

The obvious decomposition of this sentence yields:

[Somebody] removed fertile eggs from the incubator after 3 days of incubation.

The *somebody* which is needed for this transformation clearly refers to the authors of the paper, and again if we include the full title as part of the text the clarification of this zeroed *somebody* is easy to reconstruct. (There is a different kind of passive which also can occur in scientific texts and which is somewhat harder to resolve. This will be dealt with later in the discussion of transformations.) However for the purposes of the study of scientific texts it may be better and closer to the semantic intent of the text not to supply the referent for the *we* or to undo the passive. One of the assumptions underlying experimental scientific investigation is that the particular identity of the person performing the experiment is immaterial to the results of the experiment. Therefore providing the name of the author as the referent for *we* or as the referent for the *somebody* of the passive would not clarify the semantic structure of the scientific text. In fact it would make comparing the content of scientific texts much more difficult, because any sentences describing experimental procedures in a text by one author would necessarily differ from the description of procedures in a text by a different author. From the point of view of experimental scientific procedures we would like to say that the procedures were equivalent. Of course, there are texts in which the identity of the author is important to the semantic interpretation of the work. Biographies are probably the strongest example in which this is true. But it is not surprising to find that different patterns of analysis are appropriate to different kinds of texts. That this is a reasonable approach to these texts is strongly supported by the structure of the texts themselves. In all of the texts studied only two had any sentences in which *we* occurred, and even in these two texts there were only a total of three occurrences of *we*. The most common general form for a sentence with a verb of action was the passive with the second noun phrase deleted. Since this is by far the most prevalent sentence form and since, too, there seems to be good semantic justification for this form it seems best, from the point of view of normalization and from the point of view of semantic analysis, not to use the passive decomposition on these sentences.

3.2 "*It*"

The only other pronoun which occurred in any of the texts was *it*. There were two occurrences of *it* and both of them were in the same text, #068. In both cases the referent of the pronoun occurs in the preceding sentence. The sentences involved are:

Therefore, we suspected and found in rabbit serum a highly potent substance which contracts the aortic strip. In subcontractile doses, it potentiates the effect of ephedrine. A protein-free active substance, 100-1000 times more potent than the starting material, was obtained from rabbit serum by dialysis and subsequent concentration. It is stable to boiling at pH7, quite stable to boiling in strong acid, but unstable to boiling in alkali and inactivated by trypsin.

In both cases the pronoun referent is not the noun phrase in the preceding sentence which is closest to it, but rather to the noun phrase which is the subject of the underlying scientific passive sentence. There are clearly not enough cases of reference in our sample to lead us to any reliable conclusions. The absence of pronouns in our sample could be a function of length of the texts which we are dealing with, or it could be that, in fact, that pronoun reference is rare in scientific texts. From our sample it would seem that classifier words and synonyms are used much more frequently than pronouns in order to avoid boring repetition of some word or phrase.

3.3 *Ordinal pronouns*

There was one case, in text #006, of the ordinal pronoun phrase, *The latter*. The sentences in this instance were:

On refeeding, the *N*-free animals respond more poorly to a low protein (14%) diet than either the normal controls or the started group. The latter's response was superior to that of the controls.

In this case the referent is simply the last noun phrase of the previous sentence. However, the referent could be the two noun phrases conjoined by *or*. This possibility is eliminated by the occurrence of *controls* in the second sentence. Had the second sentence been:

The latter's response was superior to that of the former.

the referent for the two ordinal pronouns could be either:

The latter = either the normal controls or the starved group
The former = The *N*-free animals

or:

The latter = the starved group
The former = the normal controls

since the sentence specifies *the controls* as the second noun the referent for *the latter* must clearly be *the starved group*.

3.31 *Demonstratives*

A far more prevalent kind of interreference were the demonstrative adjectives and pronouns *this*, *that*, *these*, *those*. In the sixteen texts we looked at there were three occurrences of the demonstrative pronoun, and 13 instances of the demonstrative adjective. In all three cases of the demonstrative pronoun the referent was the whole previous text or the whole previous sentence. In the case of the demonstrative adjectives 7 of them referred to a whole sentence or set of sentences rather than to a preceding noun phrase. Thus, it seems that the demonstratives serve more to connect passages of text in scientific discourse than to avoid repetition of noun phrases. There are six cases in which the referent is to a preceding noun. In three of these the noun referred to is the noun which occurs with the demonstrative so that reconstruction of the referent is simple. In the three cases in which the referent is not the same as the noun which occurs with the demonstrative, the noun with the demonstrative is a classifier word for the referent noun. These three cases are:

guinea pigs ↔ these animals (#021)
rat epidermal preparations ↔ this tissue (#052)

glucose-6 phosphate
6-phosphogluconate
isocitrate
malate } ↔ these enzymes (#126)

The first of these cases seems an easy one to substantiate. One would expect that the classifier *animal* would be wide-spread and necessary in experimental biology papers, and in fact it does occur (#006, #015, #021, #052, #197). The second reference is substantiated in the structure of the text itself. The first sentence of the text contains the phrase *rat epidermal tissue* which allows us to establish the underlying sentence *Tissue is epidermis*. Therefore the relationship between *epidermis* and *tissue* which is necessary for the clarification of the demonstrative can be established on the basis of other sentences in the text. Thus, this is a case of inter-sentential dependency in which an ambiguity at one place in the text was resolved by the decomposition of a different sentence in the text. The last case is resolvable by elimination. In the previous two cases the referent was in the preceeding sentence; in this case there is an intervening sentence:

The method of measurement and unitage are those of Glock and McLean (1955).

However, this intervening sentence does not contain any candidate for the referent of *these enzymes* and since the next preceeding sentence is the first sentence of the text there are no other candidates.

The demonstrative in scientific texts seems to function primarily as a connecting link between parts of the text, rather than as a connective between occurrences of nouns within the text. In general, the sentence with the demonstrative contains a conclusion which can be drawn from the text to which the demonstrative refers. There is no automatic procedure for determining the extent of this reference. In this sense the whole text may be ambiguous. There may be two or more different readings for a text depending on the extent of reference. For instance in #068 we have the sentence:

These data indicate that the active substance might be a peptide, Serum Vasoactive Peptide x (SVPx).

It is not clear whether *these data* refer to the three conditions stated in the previous sentence, or to the evidence given in the whole preceding text. This seems to be an ambiguity which is resolvable only by knowing what kind of evidence is reasonable to support this conclusion, but is not resolvable on the basis of the syntactical analysis of the sentences of the text.

3.32 "*The*"

The one other kind of intersentential reference which occurs is the referential occurrences of *the*. The complexity of establishing a source for each *the* in a text discouraged the use of this technique as a general method of normalization. We would expect, for every *the N* structure in a text, to be able to establish a "*A N wh* +" source for it. This, however, is very difficult to do. There are, of course, some cases where such an underlying structure can be established, but in scientific texts there are many occurrences of *the N* where there is no previous occurrence of *N* and where *N* is not generic.

Perhaps it would be clearer if we looked at a specific text (#076). Let us look at just the first and fourth sentences of this text:

The toxic dose of rapidly acting AcS, shown as arrhythmia in the ECG is highly reproducible in the same dog on different days, providing a suitable reference for study of the effect of other drugs on the toxicity of cardiac glycosides.
The mean cardio-toxic dose of merallide indicated by widening of QRS and elevation of T wave was 0.196 ± S.E. 0.022 mg/kg, and the mean time required for half clearance of the drug from the heart was 58.3 ± S.E. 4.2 min.

The first instance of *the* is resolvable into:

The toxic dose ↔ a dose which is toxic

The source for *the ECG* is not as easy to determine. The source cannot be the generic *the* because we can have:

shown as arrhytmia in the ECG ↔ the ECG which shows arrhythmia and, as pointed out by both B. Robbins [2] and Z. Vendler,[3] the generic *the* cannot be followed by a restrictive phrase. In order to reconstruct the most likely source for this instance of the definite article we must supply semantic information about the structure of the experiment to get:

> A dose of rapidly acting AcS was given to a dog.
> An ECG was taken of a dog which was given AcS.
> An ECG which was taken ... showed arrhythmia.

From this structure we can continue the decomposition using B. Robbins schema for decomposition of relative clauses. The most likely source for *the effect* is not the generic but rather the *the* which is used as the article with unique nouns. In this context *the effect* is modifiable by restictive clauses, which eliminates the possibility of the generic:

The effect of other drugs on the toxicity of cardiac glycosides which was determined by this experiment.

If we consider that the assumption of any experimental scientist is that every experiment will result in a unique set of events which will occur every time the experiment is performed, we can understand the occurrence of the definite article with a noun which refers to the outcome of the experiment. That is, one of the sources for *the* in English is as the definite article used with a unique object, e.g. the Taj Mahal, the White House, etc. Since each experiment has a unique outcome (though there may be many instances of it) terms referring to the outcome of an experiment occur with the definite article. This seems to explain the source for the definite article in such phrases like: the effect, the result, the outcome.

It is clear that the establishment of the source for many of the definite articles in scientific texts must be founded on semantic considerations, rather than on strictly syntactic considerations.

[2] B. Robbins, "Relative Clause Adjuncts of a Noun." *Transformations and Discourse Analysis Projects* ≠47. p. 92. University of Pennsylvania.

[3] Z. Vendler, "Nominalizations." *Transformations and Discourse Analysis Projects* ≠55, pp. 7-10. University of Pennsylvania.

Though such analyses are necessary for a linguistic analysis of the texts, they do not help in the normalization of texts, nor in the determination of intersentential dependencies.

4.0 PATTERNS OF WORD REOCCURRENCE

We also looked at the patterns of word reoccurrence in some of the texts, (see Appendix III). These interrelationships were determined by noting the occurrence in every sentence of any word, except for grammatical constants, which occurred more than once in the original text. We were motivated to do this because many of the now available machine programs which claim to provide an automatic content analysis of a text are based precisely on such frequency techniques. Usually the text is first checked against a stop list and then any word which does not occur on the stop list is checked for its frequency in the text. The text used for determining frequencies is the original text. Therefore, any tacit occurrence of a word because of pro-word reference or deletion is ignored. It is our contention that even such frequency analyses would be more reliable analyses of the content of a text if all tacit occurrences of each term was made explicit. The analyses of the frequency of the words in these texts do not reveal much about the structure of the text. Moreover, the decision of what words to put on the stop list and what words to count is difficult to determine. There are some clear cases of stop words e.g. *the*, *and*. However, decisions about some classes of words are not so easy. For instance should all relational words e.g. *greater*, *increase*, *proportional*, *equal*, be on the stop list or not? They certainly are needed for a semantic interpretation of the text, yet counted in isolation it is difficult to assign meaning to them. Once the content of the stop list and the count of the words not on the stop list is completed different analyses take different routes. One technique extracts from the text that sentence in which the greatest number of most frequent words occurs, because that sentence supposedly has the densest content. Other techniques determine relative frequencies of a set of words in different parts of a text, in order to correlate difference in frequencies with differences in the focus of interest (e.g. an analysis of Russian military documents for the relative frequencies of the sets: conventional warfare-Russia vs. nuclear warfare-The United States). As can be seen from studying the results of such frequency counts as given in Appendix III compared with the more detailed results which are obtainable from a transformational analysis of the texts, a sophisticated semantic analysis requires at least some preliminary transformational analysis of the text.

5.0 DECOMPOSITION

5.1 *Coordinating Conjunctions*

The next step in the decomposition of the texts was the removal of the coordinating conjunctions. Detailed decompositions of the texts done by removing coordinating conjunctions are given in Appendix IV. The major coordinating conjunctions which occurred in the texts were *and* and *or*. There was more than one decomposition route for each of these conjunctions. In many cases the decision among these routes could be made by looking at the structure of the other sentences in the text. These decompositions, therefore, provide an interesting example of the intersentential dependencies.

5.11 "*And*"

There were four different decompositional paths taken by sentences with *and*. In two of these types deletion of like segments could occur.

5.111 *Standard "and"*

The first of these was:

$$N_1 \text{ and } N_2 \quad V \leftrightarrow N_1 \quad V, \quad N_2 \quad V$$

and:

$$N_1\ V\ N_2 \text{ and } N_3 \leftrightarrow N_1\ V\ N_2,\ N_1\ V\ N_3$$

and:

$$N_1\ V_1 \text{ and } V_2\ N_2 \leftrightarrow N_1\ V_1\ N_2,\ N_1\ V_2\ N_2$$

Examples from our texts of each of these are:

#224
The mononucleotides, thymidylic and deoxyadenylic acids appear early in the reaction and reach a plateau at about 24 hours incubation.

The mononucleotides appear early in the reaction and reach a plateau at about 24 hours incubation.

The thymidylic acids appear early in the reaction and reach a plateau at about 24 hours incubation.

The deoxyadenylic acids appear early in the reaction and reach a plateau at about 24 hours incubation.

(NOTE: There are more steps in this analysis than are shown here. The triple subject can be decomposed pairwise in two different ways, but since either decomposition yields the same sentences it is not necessary to show both paths. The range of the modifier *acids* is not extended to mononucleotides because the plural cannot act as an adjective. The choice of whether to decompose the compound noun or the compound verb first is immaterial in this case since either decomposition would yield the same set of sentences.)

#063
Coreless Kolin sine-wave electromagnetic flow meters were implanted around the roots of the aorta and pulmonary artery a week before study. ↔

Coreless Kolin sine-wave electromagnetic flow meters were implanted around the root of the aorta a week before study.

Coreless Kolin sine-wave electromagnetic flow meters were implanted around the root of the pulmonary artery a week before study.

(NOTE: It cannot be determined from this sentence alone whether the plural on *roots* is in the underlying sentences or comes in when the two underlying sentences are conjoined. There is nothing in the rest of the text either which would clarify this ambiguity.)

#068
We suspected and found in rabbit serum a highly potent substance which contracts the aortic strip. ↔
We suspected [that there was] in rabbit serum a highly potent substance which contracts the aortic strip.
We found [that there was] in rabbit serum a highly potent substance which contracts the aortic strip.

(NOTE: We will use square brackets throughout this study to enclose material which is restored.)

5.112 *Temporal "and"*
A second type of coordination with *and* requires a temporal interpretation of the sequence of sentences which are coordinated. This transformation may be represented by:

$$N_1\ V_1 \text{ and } V_2 \rightarrow N_1\ V_1 \text{ then } N_1\ V_2$$

An example of such a construction is:

#010
Antigen solutions are mixed with melted agar at 40°-50° C., allowed to gel, and quick frozen. →

Antigen solutions are mixed with melted agar at 40°-50° C.
then
Antigen solutions are allowed to gel.
then
Antigen solutions are quick frozen.

It should be noted that in the standard construction with *and* there is no fixed order to the conjoined terms. Changing the order of the terms in the conjoined phrase produces a paraphrase of the original sentence. In sentences in which terms are conjoined by the temporal *and* the order of the terms is fixed. A change in the order of the conjoined terms produces a sentence which differs in meaning from the original sentence, e.g.:

Antigen solutions are quick frozen, allowed to gel, and mixed with melted agar.

is not a paraphrase of:

Antigen solutions are mixed with melted agar, allowed to gel, and quick frozen.

while:

John bought bread, sugar, and milk.

is a paraphrase of:

John bought milk, sugar, and bread.

This temporal *and* occurs primarily with conjoined verbs, though there are some instances in which it occurs with conjoined nouns:

I like a fast game of tennis, a hot shower, and a thick steak.

In the instances in which it conjoins nouns there seems to be a strong set of implicit verbs. The sentence above might be represented as:

I like [to play] a fast game of tennis, [to take] a hot shower, and [to eat] a thick steak.

In the example given from document #010 not only is there a temporal connection between the sentences, but also the noun phrase in the second sentence is not simply *antigen solutions* but rather refers to the antigen solutions which have been treated as described in the first sentence. Perhaps a better source for this would be a set of underlying sentences with relative clauses indicating the relationship between the noun phrases, which would give:

Antigen solutions are mixed with melted agar at 40° - 50° C.

Antigen solutions which are mixed with melted agar at 40° - 50° C. are allowed to gel.

Antigen solutions which are mixed with melted agar at 40° - 50° C· which are allowed to gel are quick frozen.

All of the sentences which can undergo the temporal *and* decomposition can also undergo the standard *and* decomposition. Therefore, all of these sentences are ambiguous. The resolution of this ambiguity rests primarily on semantic grounds.

5.113 *Compound subject "and"*

There are some sentences in which *and* joins two noun phrases to form a really compound subject. That is, the two nouns are not simply joined syntactically, but rather, they refer to objects which are really conjoined. An example from our texts is:

#235
Eggs from hens fed 2.5% cottonseed oil and 7.5% corn oil contained as much stearic acid as eggs from hens fed only 2.5% cottonseed oil.

Under the standard *and* decomposition transformation this would give:

Eggs from hens fed 2.5% cottonseed oil contained as much stearic acid as eggs from hens fed only 2.5% cottonseed oil.

Eggs from hens fed 7.5% corn oil contained as much stearic acid as eggs from hens fed only 2.5% cottonseed oil.

This is obviously not the proper decomposition for the sentence. In this case there are syntactic grounds for rejecting it as a possibility. The occurrence of *only* demands some difference in the noun phrase which follows it. The following structure is not a grammatical structure:

$$(N_1\ V_1\ N_2)\text{ nom (comparative phrase) }(N_1\ V_1\text{ only }N_2)\text{ nom}$$

This, however, is not enough to distinguish these sentences in all cases, since the second noun phrase is not always identical with the first. Any sentence which contains this *and* is paraphraseable by a sentence which is identical to it except that every occurrence of this *and* is replaced by *plus*. In the text from which this example was taken this equivalent structure does occur.

Eggs produced by hens fed the basal diet plus crude cottonseed oil had larger proportions of stearic acid and smaller proportions of oleic acid than did eggs from hens fed a basal diet containing 4.1% lipide or the basal diet plus corn oil.

If we consider the sentences underlying this sentence we can obtain:

Eggs produced by hens fed the basal diet plus cottonseed oil had larger proportions of stearic acid than eggs from hens fed the basal diet plus corn oil.

Thus, we would have to say that the same text had two sentences, one from the first decomposition given and one from the latter, of the form:

$$(N_1\ V_1\ N_2)\ \text{nom (comparative phrase)}\ (N_1\ V_1\ N_3)\ \text{nom}$$

These two sentences differ only in that (comparative phrase)$_1$ is not a paraphrase of (comparative phrase)$_2$. From the point of view of scientific texts this kind of contradiction would be excluded, so that this set of sentences would be excluded. Since one of the sentences results from one of the decompositional paths of an ambiguous sentence this consideration would make us choose the alternate decomposition, and thus resolve the original ambiguity.

5.114 *"And" from rearrangement*

With some verbs which have an obligatory object, the object can be moved to before the verb and conjoined to the subject with *and*. This occurrence of *and* serves merely to distinguish between the subject and object of the verb. An example of this structure from our data is:

#063
The behavior of the right and left ventricles was similar.

This undergoes the transformation:

The behavior of the right ventricle was similar to the behavior of the left ventricle.

In general we have the transformational structure:

$$N_1 \text{ and } N_2 \text{ be} \begin{Bmatrix} \text{similar} \\ \text{different} \\ \text{equivalent} \\ \text{etc.} \end{Bmatrix} \leftrightarrow N_1 \text{ be} \begin{Bmatrix} \text{similar} \\ \text{different} \\ \text{equivalent} \\ \text{etc.} \end{Bmatrix} P\ N_2$$

5.12 "*Or*"

5.121 *Exclusive vs. Inclusive*

The major ambiguity of sentences with *or* is deciding whether the inclusive or exclusive meaning of *or* is the proper reading. If a set of terms is conjoined with *or* and the inclusive reading is intended then at least one of the underlying sentences must be true. If the exclusive reading is intended then at most one of the underlying sentences must be true. There is no syntactic means of distinguishing between these readings on the basis of the *or* sentence alone. Yet it is clear that, especially for scientific texts, the distinction between the inclusive and exclusive *or* is crucial for a proper analysis of the semantic content of the text. By looking at the structure of the other sentences in the text the distinction between inclusive and exclusive *or* can often be made. An example of each type of *or* construction, each of which is resolvable by later sentences in the text, is:

#052
Glucose oxidation by epidermal tissue may proceed through glycolysis and the Krebs cycle, the direct oxidative pathway, or via glyoxalate oxidation.

#141
The reaction between catalase and a peroxide can be formulated as shown in equations (A), (B), or (C).

(NOTE: the next sentence specifies equations (A), (B), and (C). We will not repeat them here.)

Given only the sentences themselves, a decision about which meaning of *or* is intended cannot be made. However, this decision can be made on the basis of the other sentences in the text.

5.122 *Truth-functional criteria*

In the case of these two examples the decision is based on the truth value of the underlying sentences. In the case of the exclusive *or* conjoining three terms, *A*, *B*, *C*, we know that only one of the following sets of truth values for the underlying sentences can hold:

$$\left\{\begin{matrix} \text{Sent. } A \\ \sim\text{Sent. } B \\ \sim\text{Sent. } C \end{matrix}\right\} \text{ or } \left\{\begin{matrix} \sim\text{Sent. } A \\ \text{Sent. } B \\ \sim\text{Sent. } C \end{matrix}\right\} \text{ or } \left\{\begin{matrix} \sim\text{Sent. } A \\ \sim\text{Sent. } B \\ \text{Sent. } C \end{matrix}\right\}$$

That is, only one of the underlying sentences which contain the conjoined terms can be true. In the case of the inclusive *or* conjoining three terms, *A*, *B*, *C*, we know that the only possible set of truth values which could not hold is:

$$\left\{\begin{matrix} \sim\text{Sent. } A \\ \sim\text{Sent. } B \\ \sim\text{Sent. } C \end{matrix}\right\}$$

The following sentence appears in text #052:

Evidence has been adduced to support these various possibilities.

By making explicit the demonstrative we get:

Evidence has been adduced to support the possibility that glucose oxidation by epidermal tissue proceeds through glycolysis and the Krebs cycle.

Evidence has been adduced to support the possibility that glucose oxidation proceeds through the direct oxidative pathway.

Evidence has been adduced to support the possibility that glucose oxidation proceeds via glyoxalate oxidation.

If we consider the paraphrastic relationship in scientific texts:

$$\text{There is evidence to support } X \xleftrightarrow{\text{PARA.}} X \text{ is true}$$

we can conclude from the second sentence from text #052 the following truth values for the underlying conjoined sentences:

$$\left\{\begin{matrix} \text{Sent. } A \\ \text{Sent. } B \\ \text{Sent. } C \end{matrix}\right\}$$

This means that the *or* in the first sentence of the text must be the inclusive *or*. The following sentence from text #141 allows us to resolve the ambiguity of *or* in that text:

The reversibility of the reaction between catalase + CH_3OOH eliminated formulation (*C*) and a detailed study of the equilibrium was consistent with (*A*) but not with (*B*).

From this sentence we can conclude:

$$\left\{\begin{array}{r} \text{Sent. } A \\ \sim\text{Sent. } B \\ \sim\text{Sent. } C \end{array}\right\}$$

which would reinforce the exclusive *or* reading of the first sentence.

5.123 *Structural criteria*

The distinction between inclusive and exclusive *or* in the above examples was not based on purely syntactic grounds, but rather, involved considerations of paraphrase and truth values of sentences. There are some cases, however, when the decision can be made on syntactic grounds. Consider the sentence:

#006

Birds depleted of 1/3 their initial body weight, either by starvation or by feeding an *N*-free diet, show marked differences in carcass composition at equal body weights.

This yields:

Birds were depleted of 1/3 of their initial body weight, either by starvation or by feeding an *N*-free diet.

The birds show marked differences in carcass composition at equal body weights. ↔

[some] birds differed from [other] birds.

If we read the first sentence with the exclusive *or* we get two sets of readings for the underlying sentences:

1. { Birds were depleted of 1/3 their initial body weight by starvation.
   Birds were not depleted of 1/3 their initial body weight by feeding an *N*-free diet. }

2. { Birds were not depleted of 1/3 their initial body weight by starvation.
   Birds were depleted of 1/3 their initial body weight by feeding an *N*-free diet. }

If we use the inclusive *or* reading we get the two sets above and an additional set:

{ [Some] birds were depleted of 1/3 their initial body weight by starvation.
[Other] birds were depleted of 1/3 their initial body weight by feeding an *N*-free diet. }

In order to have a reading for the underlying sentences:

[Some] birds differed from [other] birds.

we must choose the inclusive reading of *or*. This choice, made on structural grounds here, is reinforced later in the text by the occurrence of both:

N-free birds
starved birds.

6.0 PASSIVES

When we were discussing the problem of the referent of the pronoun, *we*, we mentioned that there were different kinds of passives in these texts. It is important to point out the distinction between these structures because the difference between them will be important for the structural description of text and for the subsequent semantic analysis of these texts. Consider the following three sentences:

#193
Excised, curarized frog sartorius muscles were massively stimulated in 1 mM caffeine-Ringer's at 24 C.

#006
For the amino acids methionine and lysine the absolute requirements were increased for the depleted animals.

#010
Sections are made at 32 mu in a cryostat at −15 C.

All of these sentences are ambiguous. For two of the sentences (#193, #010) there are three possible readings:

N_1 be V_1 en P $N_2 \leftrightarrow N_2$ V_1 N_1
N_1 be V_1 en P $N_2 \leftrightarrow$ [the experimenter] V_1 N_1 P N_2
N_1 be V_1 en P $N_2 \leftrightarrow$,[forces of nature] V_1 N_1 P N_2,

There are two different transformational explanations of the third possibility. In the first one given, the subject of the transformed sentence is *the forces of nature* or some other such general noun phrase. In the second the verb is simply treated as the compound past tense. The decision between these two formulations depends on other considerations in the form of the particular transformational grammar used, but this choice is not crucial to us here. In either case there are three different decompositions for the two sentences in question. For the third example (#006) only the latter two decompositions are possibilities. The first transformational possibility can, in fact, occur with prepositions other than *by*, e.g.:

Desperation is bred in poverty. ↔ Poverty breeds desperation.
Strength is developed through excercise. ↔ Exercise develops strength.

The three possible readings for the sentence from #193, in abbreviated form, are:

Caffeine-Ringer's stimulated muscles.
[The experimenter] stimulated muscles in caffeine-Ringer's.
[The forces of nature] stimulated muscles in caffeine-Ringer's.

From the sentence in isolation this ambiguity cannot be resolved, but when we consider the next sentence of the text:

This caffeine concentration raised twitch tension output to 190% of normal Ringer's values.

we have strong evidence to choose the first formulation. That is, since it is the caffeine concentration which raised the twitch potential it is reasonable to suppose that it was the caffeine concentration which stimulated the muscles.

Establishing any kind of criteria to distinguish the preferred reading of the other two sentences is difficult. Yet in the context of the text they are not ambiguous. Semantic arguments can be used to distinguish between the two readings in the case of #010. That is, we know that, in fact, experimenters make sections and that sections are not produced by the forces of nature, but such a resolution of the ambiguity does not give us any useful way of distinguishing between the readings in other sentences. However, there are certain patterns of sentence structures in texts which would lead us to choose one reading over the other. We will go into the details of these patterns in the next chapter. As a brief example of these kinds of dependencies we can point out that if the preferred reading is the one in which *the experimenter* occurs as the subject of the underlying sentence, then there will be another sentence in the text in which N_1 in some other situation is contrasted to N_1 in the situation described by N_1 V_1 N_2. In more intuitive terms, the occurrence of a sentence in which the experimenter performs some action on some system will be accompanied by a sentence in which the results of this action will be contrasted to some other state of the system. This other sentence can be described in terms of repeated occurrences of N_1, V_1, and N_2, and the pattern can, in this way, be made explicit.

7.0 COMPARATIVE STRUCTURES

One other important kind of structure which occurs with great frequency in scientific texts is the structure we will refer to as the comparative structure. Though the semantic range of this structure is clear, it cannot be represented by a single syntactic structure. The kinds of structures which we will try to delimit here all involve specifying a change from one state to another. In most scientific texts the degree of change is quantifiable. Examples from our texts of such structures are:

#006
For the amino acids methionine and lysine the absolute requirements were increased for the depleted animals.

#047
Peritoneal fluid osmolality fell.

#235
Eggs from hens fed 10% cottonseed oil contained more stearic acid than eggs from hens fed 2.5% cottonseed oil.

#018
At hatching, the mean weight of the bursa for 20 F_1 chicks from the TP and EA groups were 31 $\pm$ 7 and 43 $\pm$ 7 mg., respectively. The differences were highly significant.

The two major structures which are involved in such comparatives are:

1. $$N_1 \text{ be} \begin{Bmatrix} A \text{ er} \\ \text{more } A \\ \text{similar} \end{Bmatrix} \begin{Bmatrix} \text{than} \\ \text{to} \end{Bmatrix} N_2$$

2. $N_1\ V_{\text{meas.}}\ \{\text{from } N_2\ \{\text{to } N_3\}\}$

where $V_{\text{meas.}}$ = change, increase, decrease, equal, differ

N_2 and N_3 in structure 2 may be numbers

When the second two noun phrases in structure 2 do not appear, as in #006, then there is an understood structure:

$N_1\ V_{\text{meas.}}$ [from the normal (or previous) state]

In these texts there is a paraphrastic structure which is often needed to explicate such structures:

controls $\xleftrightarrow{\text{PARA.}}$ normal (untreated) state

This relationship is not made explicit in any of the texts, but rather is an assumed part of the semantic knowledge of the reader. In general there is no definition of *controls*, yet the results of measurements are often compared to the results of measurement of *controls*. Therefore this paraphrastic equivalence must be assumed in order to interpret the structure of the comparisons in the text.

7.1 *Structure of noun phrases*

In the type of comparative phrases which we have discussed, the noun phrases in the structures most often differ in only one item, e.g. in #235 and #018, the two groups being compared differ in only one respect. There are, however, some instances when there is no explicit semantic similarity between the two nouns, e.g.:

#063
Peak flow, peak acceleration, and ejection time all changed in proportion to the change in stroke volume.

In this case all three of the comparative structures which result from decomposition of the *and* have non-matching noun phrases. However, a previous sentence in the text establishes that these nouns are indeed related:

During marked spontaneous sinus arrhythmia, we measured the effects of variations in heart rate on the stroke volume, peak flow, peak acceleration and ejection time of the right and left ventricles of conscious dogs.

Thus, the two noun phrases in such comparative structures either are explicitly similar, or are established as similar by some previous sentence in the text. The one other type of relationship between the noun phrases in these structures is the one we mentioned previously, i.e. the second noun may be missing in which case it is understood to be [N_1 in its normal (or previous or untreated) state]. Or the second noun may be *controls* in which case it will have not have been previously introduced in the text as related to the noun in question, but rather is to be regarded as necessarily related to N_1 and synonomous with [N_1 in its normal (or previous or untreated) state].

8.0 SUMMARY

As can be seen by the decompositions of the texts given in Appendix IV certain patterns of repeated syntactical forms which are obscured in the original text become much clearer when we normalize the text. And, too, certain ambiguities which arise when we decompose a single sentence in the text can be resolved when the intersentential relationships are taken into account. As we will see in the next chapter, once we have established a syntactic description of a text we can go on to show that certain patterns of syntactic descriptions reoccur in many texts. In this way we can use overall patterns of sentence forms to distinguish among types of texts. These syntactic patterns can be associated with the semantic content of the text. The fact that these patterns of sentence structure reflect semantic patterns will allow us to make suggestions about procedures for extracting information from these different kinds of texts. In chapter 4 we will try to indicate some of the ramifications the discovery of these various patterns has on the storage, retrieval, and representation of scientific texts.

3

SYNTACTIC STRUCTURES IN TEXTS

1.0 DEFINITION OF CRITERIA FOR STRUCTURE

In this chapter we will discuss certain patterns of syntactic structures which appeared in the scientific texts with which we were dealing, and, since we assume our sample to be a representative sample, which are typical patterns of scientific texts. There are three different aspects of syntactic patterns which could be studied. The first of these is a study of which transformations actually did occur in the empirical data, but since any sentence which is grammatical may appear in some text, every structure found in any sentence must occur in some text. From the point of view of designing real programs to handle real texts, however, such a study would not be without its value. It may be that certain transformations which are necessary for a total grammar of English may not occur in scientific texts, and may therefore not be necessary for an adequate grammar of scientific English. Eliminating such superfluous transformations would simplify the grammar and perhaps eliminate some ambiguous structures. However we will not deal here with any such data.

A second aspect of the concept of syntactic patterns in texts is the influence such structures have on one another. We have discussed certain instances of such intersentential dependencies in the previous chapter and will discuss certain other such dependencies in this chapter. However, the explication and resolution of this type of dependency is only a necessary preliminary to the real study. In order to describe the general syntactic patterns of texts it is first necessary to assign a syntactic description to each of the texts individually. The resolution of intersentential dependencies will allow us to decide among various possible readings of an individual text, and will therefore assign a description to each text which is not simply an aggregate of all of the structures assigned to each of the sentences individually, but is rather the result of the interaction of these separate descriptions.

The third aspect of syntactic patterns, and the one which will be dealt with at length in this chapter, is the description of the different general patterns of transformations and cooccurrences which delimit a type of text. The fact that such different types of texts exist is neither necessary nor obvious. That is, there is no reason to suppose that there would be types of texts which could be described by the order and occurrence of transformations. Moreover, these different types of texts reflect not simply the stylistic variations among texts, but also they mirror certain semantic differences among the texts. In short, different kinds of messages are encoded in different patterns of transformations. In this chapter we will establish general criteria for the description of such patterns and present the evidence for the establishment of three major types of scientific texts. In the next chapter we will discuss the ramifications which the distinctions between types of texts have for text representation, storage, and retrieval.

Before describing any of the different patterns of structures which can serve to distinguish among different types of texts it is important to make explicit the criteria upon which any description of a type of text should be based. We will be using primarily syntactic criteria to distinguish among different types of texts. One of the distinguishing characteristics of textual types will be the occurrence of certain transformations. Another of the distinguishing characteristics will be the pattern of repetition of noun phrases and verbs phrases within the text. In some cases, especially in the case of comparative structures which were mentioned at the end of the previous chapter, it will be best to establish a normalized syntactic form to describe a set of syntactically different structures. This will be done by establishing that for every sentence in one of the variable forms there is a sentence in the normalized form which is a paraphrase of it. Therefore the normalized form can be used to represent the variable forms in order to make the description of the structural patterns more uniform. The sentences which appear in the real text and are the result of transformations which change the sentence from the normal form to the variable form can be regarded as stylistic variants.

Each type of text will be represented by a set of syntactic structures by which it is characterized. Some of these structures will be obligatory and other of the structures will be optional. We will illustrate these various patterns by applying them to the body of data with which we are primarily concerned. In some cases it will be possible to say that certain sentences which are necessary for the completion of the text are missing, and that without these sentences the set of sentences is a deficient text. In other cases it will be possible to point to some sentences which occur in the real data, but which, in fact, could be eliminated without affecting the texthood of the set of sentences. These "extra" sentences, as we will see when we look at the texts involved, are primarily sentences which cite other work done in the field.

1.1 *Occurrence of transformations*

One of the primary distinguishing characteristics of types of texts will be the occurrence of certain key transformations. That is, the set of transformations which are used to encode the semantic information in a text is not completely arbitrary. Therefore, the fact that a particular transformation is needed in order to describe the syntactic structure can be significant for the later semantic analysis of the text. However, not every transformation plays a role in distinguishing among textual types. For instance, the standard use of the coordinating conjunction *and* occurs in almost all texts and, in general, does not play a critical role in specifying the type of the text. On the other hand, the temporal conjunction *and* does serve to distinguish among types. This distinction may allow us to distinguish between the readings of a sentence with *and*. Thus, if a text exhibits a structure which is not that in which the temporal *and* occurs, then the other reading of *and* is the preferred reading.

1.2 *Position in text*

The structure of transformations needed to describe different types of texts must not only specify the particular transformations which can occur, but also the order in which they occur. That is, there are restrictions not only on the transformations which occur, but also on their relative order. A trivial example of these order restrictions is the fact that pro-words do not occur in the first sentence of a text. This restriction, however, will hold for all scientific texts and will not be an order restriction which will distinguish among types. A more pertinent example of an order restriction which does distinguish among textual types is the occurrence of a container sentence as the last sentence of a text. A container sentence is a sentence with the structure:

N V that Sent.

where the verbs that can occur are from a special class of container verbs such as, shows, proves, indicates, etc. In the texts we studied more than 1/3 of the texts ended with a container sentence. In all but two of these cases certain other special structural features occurred in the text also. Both the two aberrant cases and the other special features will be discussed at length later in this chapter.

The position of a transformation in a text may not only be defined in relation to the structure of the whole text, but also in relation to other transformations in the text. Thus, we can specify that one structure must be preceeded by some other structure, though the absolute position in the text of either feature is not specified.

1.3 *Non-occurrence of transformations*

In some cases a significant criteria to distinguish among textual types will be the non-occurrence of some structure. It may be that the number of structures which could occur is very large, but that it is significant that some particular structure does not occur. In that case the pattern could be specified by making explicit the form which cannot occur rather than those which can.

1.4 *Interrelationships among transformations*

Thus far we have established as criteria for specifying the structure of texts

a. the occurrence of specific transformations
b. the position in the text where these transformations occur
c. the non-occurrence of some transformations.

Another criteria for specifying the structure is making explicit the dependencies among the transformations in the text. Since all of the obligatory transformations which are the basis of the description of the text must occur together in the text, they have a kind of mutual dependency. However, there can be less broad dependencies among the transformations of a text. It may be that only a certain subset of transformations in the text are interdependent. That is, the occurrence of some particular transformation at a point in the text in which only some general structure is required by the overall pattern of the text may affect other structures in the text. For instance, the first sentence of one form of text is specified simply as:

NP_1 be Ven NP_2

Yet if this sentence, in fact, has any of the forms:

NP_1 and NP_2 be Ven NP_3
NP_1 be Ven$_1$ and Ven$_2$ NP_2
NP_1 be Ven NP_2 and NP_3

then the structure of many of the succeeding sentences takes a form which is somewhat different from that specified in the more general pattern. In this way, we can establish subclasses of patterns of texts. The larger class will be adequately described by the more general structure. The subclasses are established by specifying more restricted descriptions of some of the structures in the text. This will be of importance to us in describing the most prevalent pattern in the texts we considered.

There will be some interdependencies which will not be useful in differentiating among types of texts. However since they are important in describing the structure of the text they will be pointed out when they occur. These dependencies are essentially stylistic in nature and will therefore not have any ramifications in subsequent semantic interpretations of the text.

1.5 *Interrelationships among noun phrases and verb phrases*

The structure of a text is not only determined by the transformations which occur, but also by the relationships among the noun phrases and verb phrases which occur in sentences of the text. Many sentences in a text may have the same subject noun phrase and differ in the verb phrase which occurs with this noun phrase, or the same verb phrase may reoccur with varying noun phrases. In some cases the subject noun phrase of one sentence may occur as the object noun phrase of another sentence. These patterns of reoccurrence establish relationships among the sentences of a text. The difference in the pattern of reoccurrence can serve to distinguish among various types of texts. In certain kinds of texts the same noun phrase will be the subject of many sentences in the text. In other types of texts no two sentences will have the same noun phrase as subject. In general, the pattern of noun phrase reocurrence will be more significant in distinguishing among types of scientific texts than will be the reoccurrence pattern of verb phrases.

In discussing the pattern of reoccurrence we will always be considering normalized texts and not original texts. We will therefore consider the clarification of a pro-word referent to be a reoccurrence of a phrase. In scientific texts it will also be necessary to consider certain classifier words to be repetitions of the more specific phrase. For example, we will consider the phrase "N-free animals" to be a reoccurrence of the phrase "N-free birds" (#006).

Establishing the proper interconnections among generic and specific terms is a problem of the lexicon. It is to be expected that certain classifier relationships which function for one scientific discipline will not work for another discipline.

2.0 MAJOR STRUCTURES

As we have mentioned above there are three major types of scientific texts which we will discuss in detail in this chapter. There may, of course, be other types which we have not encountered in our sample. However, the methods of analysis and description established in this study can be extended to include new types of texts as they are encountered. All of the texts in our sample, except one, are examples of one of these three types. The one exception, #208, "Effect of Technique on the Shape of the Salivary Curve", seems to be more discursive than scientific. It discusses the sensitivity of the salivary curve, but does not evaluate this sensitivity by any kind of measure. Nor does it supply any data on the use of this technique in any experimental situation. From an intuitive point of view it does not seem to be the same kind of rigorous scientific document as are the other texts in the sample. Therefore, since our formal criteria also excludes it we will have to say that from the standpoint of this study it is not a scientific document.

All of the texts in this sample are particularly brief texts. All but the one mentioned above conform to one of the major types of text. However, it may be that in longer texts more than one pattern can occur in the same text. One would expect that different portions of the longer text would exemplify different types of text. A more complex text could then be described as a succession of shorter texts, each conforming one to of the general types described in this study. Describing longer texts in terms of the shorter texts of different types would affect the treatment of these texts in storage and retrieval systems. Since we will show that different types of texts require different kinds of analysis to represent their semantic content, any text which is described in terms of more than one of these types will also require a differing semantic analysis.

2.1 *Controlled experiment*

The most prevalent type of text in our sample was the type which we will call "controlled experiment". This type of text and the other types of texts will be described in terms of the criteria discussed above. The name of the type reflects the kind of semantic information carried by the structure of the text, but the text is not defined in terms of this semantic content, but rather in terms of formal criteria. In this sense the name of the type of text is simply a mnemonic device.

Thirteen of the twenty-eight texts in the sample data were of the controlled experiment type. These can be subdivided into three subtypes which we will call by the mnemonic names of:

a) treated group vs. normal state
b) group 1 vs. group 2
c) treatment 1 vs. treatment 2.

These subtypes will also be defined in terms of the formal criteria established above. An important characteristic of texts of this type is the occurrence of comparative structures throughout the text. The difference between type 1a and types 1b and 1c can be established on the basis of the zeroings which occur with these comparatives. Type 1b and 1c are somewhat more difficult to differentiate because once the two treatments which the single original group undergoes is established in the text the original single group becomes differentiated into two groups. Once these two groups are established the form of the text becomes similar to that of 1b, in which the two groups are specified at the beginning.

2.2 *Hypothesis verification*

Ten texts in our sample were of the type we will call "hypothesis verification". Of these ten, eight were of the subtype, 2a, in which a single original hypothesis is examined. The results of this examination can be either to confirm or disconfirm the original hypothesis. In all eight cases the hypothesis in question appeared in some other work which is cited in the first sentence of the text. The two texts of subtype 2b treated several explanations of the same phenomenon, and tried to decide among them. In one case, #141, the explanations were mutually exclusive. In the other case there was evidence that all three possible explanations played a part in the process, but the primary explanation was still undetermined.

2.3 *Technique-descriptive*

In this type of text there is little hypothesizing and no control of the process. These texts primarily describe a scientific technique or phenomenon. Few conclusions are drawn from the information and very little comparison with any other information is done. It is usually pointed out in these texts that the technique is being applied to some new data, or that the description of the phenomenon has not yet been provided.

3.0 CONTROLLED EXPERIMENT

As we mentioned above the three major types of substructures in this category are,

a) treated group vs. normal state
b) group 1 vs. group 2
c) treatment 1 vs. treatment 2.

The texts which belong to these types are:

1a	1b	1c
#047	#097	#006
#063		#018
#076		#021
#109		#096
#159		#235
#193		
#259		

It is surprising that there are not more texts in our sample which contrast the behavior of two or more different kinds of sample populations. It may be that while this is the case in experimental biology, other scientific fields will have more texts of this type. It would seem that the primary procedures in experimental biology as reported in the Federation Proceedings involve studying the effect of a drug or procedure on a single type of organism or studying the effects of two or more different kinds of treatment on organisms of the same type. In informal terms all of these texts state the variables which the text will treat in the first sentence, followed by a number of sentences in which certain properties of these variables are measured and evaluated, and ending with a sentence which states a theoretical conclusion which can be drawn from the data.

In more formal terms this can be represented by the series of sentence structures:

N_1 be Ven
$\{N_1$ in respect X_1 be V_{measure} by method $Y_1\}$
N_1 in respect X_1 be $>$ or $<$ or $=$ normal state

.

.

.

$\{N_k$ in respect X_k be V_{measure} by method $Y_k\}$
N_k in respect X_k be $>$ or $<$ or $=$ normal state
{The experiment} $V_{\text{container}}$ that $S_{\text{conclusion}}$

(The braces enclose optional material.) In many texts the method by which the measurements of a particular characteristic are determined is not specified in the text. However, in the cases in which the method is not specified some standard method is assumed. In some texts instead of *the experiment* a pro-word like *this* occurs. The number of comparative sentences which can occur in a text can vary. In one text in our sample, #021, there are no comparative sentences. This text is the shortest text in the sample and is only eighty words long. There is a sentence in the text which says:

Sera from guinea pigs were tested for agglutinins and complement-fixing antibodies after vaccination and again following challenge.

Unfortunately the result of this testing is not reported and all we get is the conclusion that:

Phase I preparations were shown to be superior to Phase II preparations in protective potency, although not necessarily in the elicitations of CF antibodies.

In terms of the criteria we have established this text is a deficient text. In order to become an acceptable text sentences of the form:

Sera from guinea pigs in respect to agglutinins were > or < or =normal state.

Sera from guinea pigs in respect to complement fixing antibodies were > or < or = normal state.

must appear in the text. Our criteria, therefore, are stringent enough to discover aberrant texts in our body of data and, furthermore, explicit enough to allow us to complete the text from the structure of the rest of the text.

Though the structure of the controlled experiment text given above is general enough to cover all cases of this type of text, the structure of the subtypes, 1b and 1c, can be made more explicit. The structure for 1b can be represented as:

N_1 be Ven
N_2 be Ven
$\left\{\begin{array}{l} N_1 \text{ in respect } X_1 \text{ be } V_{\text{measure}} \text{ by method } Y_1 \\ N_2 \text{ in respect } X_1 \text{ be } V_{\text{measure}} \text{ by method } Y_1 \end{array}\right\}$
N_1 in respect X_1 be > or < or = N_2 in respect X_1

.
.
.

$\left\{\begin{array}{l} N_1 \text{ in respect } X_k \text{ be } V_{\text{measure}} \text{ by method } Y_k \\ N_2 \text{ in respect } X_k \text{ be } V_{\text{measure}} \text{ by method } Y_k \end{array}\right\}$
N_1 in respect X_k be > or < or = N_2 in respect X_k
{The experiment} $V_{\text{container}}$ that $S_{\text{conclusion}}$

3.1 *Establishment of classes*

The structure for the subtype 1c differs from the above structure only in the first two lines. In type 1c two different treatments are performed on members of the same group. This different treatment of members of the same group results in two different experimental groups. The structure of the text once these two groups are established is identical with the structure of the type in which two groups are established at the outset. The first two lines of type 1c will be of the form:

[Some members of] N_1 be Ven_1
[Other members of] N_1 be Ven_2

Some examples of this structure from the texts are:

#006
Birds were depleted of 1/3 their initial body weight by starvation.
Birds were depleted of 1/3 their initial body weight by feeding a *N*-free diet.

#018
Fertile eggs were removed from the incubator after 3 days of incubation and dipped in 670 mg of testosterone propionate per 100 ml of ethyl alcohol (TP).
Fertile eggs were removed from the incubator after 3 days of incubation and dipped in ethyl alcohol (EA).

It should be noted that the preferred reading of these sentences in scientific texts is paraphrasable by the structures given above which include the phrases: *some members of* and *other members of*. None of these sentences are paraphrasable by:

N_1 be Ven_1
[The same] N_1 be Ven_2

Both of the sentence structures from which the normalized pairs of sentences given in the example were derived are ambiguous. These ambiguous structures are:

#006
N_1 be Ven_1 or Ven_2

#018
N_1 be Ven_1 and Ven_2

As we have pointed out in Chapter 2 there are other structures in the text which require that the reading which we have called the preferred reading of these sentences be the structure chosen. The structures which mitigate in favor of the preferred reading are the comparative structures which appear later in the text. We will discuss these structures in detail a bit later in this chapter.

Once the two groups have been established by the sentences which vary in the verb phrase, there is often a set of transformations by means of which the differences between the verb phrases becomes adjectivized and is used to modify the original noun. At this point, there are two different noun phrases which occur in the comparative structures and the subtype 1c behaves just like the subtype 1b in which the noun phrases differed from the beginning. In our examples we get:

#006
The *N*-free birds respond more poorly to a low protein (14%) diet than the starved birds.

#018
The mean weight of the bursa for 20 F_1 chicks from the TP group was 31 ± 7.

The mean weight of the bursa for 20 F_1 chicks from the EA group was 43 ± 7.

The differences were highly significant.

3.2 *Parenthetical structures*

The two examples above both illustrate a common scientific parenthetical structure. Sentences with this structure have within them a phrase enclosed within parentheses. This phrase does not function in the syntax of the sentence and therefore if it is removed from the original sentence the resulting sentence is grammatical. There are three major sources for these parenthetical phrases:

1. CITATION

$N_{\text{proper name of person}}$ says S in $N_{\text{proper name of book}}$
$\leftrightarrow S\ (N_{\text{person}}, N_{\text{book}})$

e.g. #282
The I to II transformation involves reduction in cobalt valence, loss of cyanide and incorporation into II of an adenosine moiety, derived from ATP (Weissbach *et al.*, J. Biol. Chem. 236, PC 40, 1961.)

2. ABBREVIATION

"X" will be replaced by "Y" $\leftrightarrow X\ (Y)$

e.g. #193
The times from stimulus to onset of latency relaxation (L_R) showed no change compared to these times in normal Ringer's. ... Potentiating NO_3-Ringer's shortens L_R.

3. SPECIFICATION

The value of N_1 is equivalent to $N_2 \leftrightarrow$
$N_1\ (N_2)$

e.g. #063
These findings indicate that over a wide range of diastolic filling times (0.3 to 0.8 sec.) during a steady state in a conscious dog all the characteristics of ventricular contraction which we have measured change with the length of the diastolic filling period.

The parenthetical structures which come from source 3, specification, do not play a major role in delimiting the structure of a type of text. However, the other two sources for this structure, citation and abbreviation, are important in differentiating among types of texts. The occurrence of some form of citation in the first sentence of a text is one of the marks of a hypothesis supporting structure. There are other kinds of citation structures, but the parenthetical form is a very common one. The second source for the parenthetical structure, abbreviation, is used to make clear the patterns of repetition of phrases throughout the text. This set of transformations will allow us to say that the second sentence in the example from #193 given above is equivalent to the sentence:

Potentiating NO_3-Ringer's shortens the time from stimulus to onset of latency.

In order to differentiate among the various possible sources for the parenthetical structure we can look at the type of material enclosed by the parenthesis, and we can consider the structure of the text. If the source for the parenthetical structure is 3, abbreviation, then there will be a succeeding sentence in which the term enclosed by parentheses will appear not enclosed by parentheses. If the term which we have designated as *X* in structure given above is substituted for this occurrence of the term designated as *Y*, the resulting sentence will be grammatical. Moreover, in terms of later analysis of the text this substitution form is really the sentence which should occur for any analysis.

If the source for the parenthetical expression is either of the other two possibilities, citation or specification, then there will not be a succeeding sentence with this structure. These two possibilities can be differentiated because of the form of the term enclosed by the parentheses. The citation form will always have a proper noun with the parentheses while the specification form will usually have a numeral. The disambiguation of this parenthetical structure is one more instance of intersentential dependencies in which the structure of one sentence in the text is determined by the structure of another sentence. Moreover, this is an instance of a non-trivial restriction in the order in which structures may occur in a text.

3.3 *Measurement statement*

The measurement statement in the structure of type 1, controlled experiment texts, does not always occur explicitly in the text. In a sense it is not really correct to say that it is an optional structure, because if it does not occur explicitly it can always be reconstructed from the rest of the structure of the text and the new text which includes this reconstructed sentence will be a paraphrase of the original text. As we will see in the examples which will be given below these reconstructed measure sentences are, at the level of text, very much like reconstructed zeroed material at the level of sentence.

First let us consider those instances in which an explicit measure sentence does occur:

#197
Expiratory volume [of cats] was measured with a Wedge spirometer.

In terms of our structure this would be represented as:

Cats in respect of their expiratory volume were measured by method of a Wedge spirometer.

This second sentence is a normalized paraphrase of the sentence which occurs in the original text. In some cases the sentence which would count as the measure sentence does not occur in the usual way we would expect. For instance, we find the following normalized sentences:

63
Coreless Kolin sine-wave electromagnetic flow meters were implanted around the roots of the aorta.
Coreless Kolin sine-wave electromagnetic flow meters were implanted around the roots of the pulmonary artery.

The preceding sentence in the text is:

63
During marked spontaneous sinus arrhythmia, we measured the effects of variations in heart rate on the stroke volume peak flow, peak acceleration and ejection time on the right and left ventricles of conscious dogs.

According to our normalization procedure we will first make this a measure sentence and then remove the coordinating conjunctions. This would give us eight partial measure sentences without the *by method Y* phrase. They can be represented by:

The effects of variations in heart rate on the

$$\left.\begin{array}{l}\text{stroke volume}\\ \text{peak flow}\\ \text{peak acceleration}\\ \text{ejection time}\end{array}\right\}\text{ of the }\left\{\begin{array}{l}\text{left}\\ \text{right}\end{array}\right\}\text{ ventricle of conscious}$$

Dogs was measured during marked spontaneous sinus arrhythmia.

We have seen from the previous example the *by method Y* phrase is signalled by the mention of some measuring instrument. We can adjust our original schema to read:

$$N_1 \text{ in respect } X \text{ was measured with } N_{\text{measuring instrument}}$$

In scientific texts nouns signifying measuring instruments most often end with *meter* or *graph*. With this slight amendment in mind we can derive the *by method Y* addition to the eight partial measure sentences represented above:

...with coreless Kolin sine-wave electromagnetic flow meters which were implanted around the roots of $\left\{\begin{array}{l}\text{the aorta}\\\text{the pulmonary artery}\end{array}\right\}$ a week before study.

The binary choice in the measuring instruments between that implanted at the root of the aorta and that implanted at the root of the pulmonary artery corresponds to the binary choice of left ventricle and right ventricle in what is being measured.

In many texts there is no explicit mention made of either a measuring instrument or of any measuring being done. We find instead sentences such as:

#018
The mean weight of the bursa of 20 chicks was 31 $\pm$ 7.

#047
Plasma osmality [of dogs] rose initially then fell.

Sentences of this type allow us to reconstruct the zeroed sentences:

#018
Chicks in respect to the bursa were measured by weighing.

#047
Dogs in respect to plasma osmality were measured by the standard technique.

These sentences are implicit in the text. In any text the occurrence of a sentence in which a precise or a relative measurement is stated allows us to reconstruct a sentence in which this measurement is determined. If no method of ascertaining this measure is given in the text one can assume that some standard method was used. This is analogous to the sentence level restorations of zeroings. Given:

John ate.

we can reconstruct from this zeroed form the full sentence:

John ate food.

The reconstructed zeroed portion of a text or sentence is generally non-specific and refers to the standard phrase which would occur in this position with the other given portions of the text. If some non-standard phrase is meant to occur in this position then the phrase will not be zeroed, but rather will be made explicit.

3.4 *Comparative structures*

As we have pointed out above there is a close relationship between measurement sentences and comparative sentences. Given a comparative sentence we can reconstruct the zeroed measurement sentence, but given a measurement sentence we cannot reconstruct a zeroed comparative sentence. The kind of structure which we have here been referring to as a measurement sentence should be kept distinct from a sentence in which the results of some measurement sentence are given. An instance of such a sentence is:

#010
Sections are made at 32 mu in a cryostat at $-15°$ C.

This in itself is not a measurement sentence, but a zeroed measurement sentence can be reconstructed to give:

Sections with respect to temperature were measured by a thermometer.

We have not previously mentioned these sentences, which we will call measurement report sentences, because they do not occur singly in the controlled experiment structure which we are now discussing. They do, however, occur in the technique-descriptive texts and we will discuss them in that section.

As can be seen from the structural description of the controlled experiment text given above, (p. 55, 56) comparative sentences play a major role in these texts. Texts of subtype 1a differ from the other two subtypes in the form of the comparative sentences which occur within them. The more explicit form of the comparative occurs in subtypes 1b and 1c. In these texts we find the standard structure:

N_1 V Aer than N_2

#097
Activity of female suspensions is 80% greater than male suspensions at pH 8.7.

if the form Aer is not an acceptable form we have the variant:

N_1 V more A ly than N_2

#006
The N-free animals respond more poorly to a low protein diet than the starved animals.

In some cases the comparative structure is very similar to the ones above, but the form of the comparative term is more fixed than those indicated above. There are very many instances of the structure:

$$N_1 \text{ be } \left\{ \begin{array}{l} \text{different} \\ \text{similar} \\ \text{equivalent} \\ \text{superior} \\ \text{inferior} \\ \text{etc.} \end{array} \right\} \begin{array}{l} \text{to} \\ \text{from} \end{array} \quad N_2$$

#006
The starved birds response was superior to that of the controls.

In order to normalize all of the different comparative structures given above as well as some other types which will be discussed below (e.g. #197 The arterial pH rose.) we will postulate a fundamental form of the scientific comparative. Every sentence which we will call a comparative will be a paraphrase of a set of sentences of this fundamental form. There are some segments of the fundamental set of sentences which can be zeroed during the process of transformation which changes this set of sentences to the actual comparative which occurs in the text. The set of sentence forms which we will call the fundamental scientific comparative form appears explicitly in one text:

#018
The mean weight of the bursa of 20 F_1 chicks from the TP group was 31 ± 7 mg.
The mean weight of the bursa of 20 F_1 chicks from the EA group was 43 ± 7 mg.
The difference was highly significant.

In informal terms every scientific comparative involves the determination of a value in terms of some scalar measure, the comparison of this value with some other determined value in the same scalar measure, and the determination of the significance of the difference between them. In any scientific text the mere fact that the numerical values assigned to two items are not identical does not mean that the difference between them is significant. The determination that two items do indeed differ from one another can only be made by knowing the extent of the uncertainty of the measure which is introduced by the design of the experiment and the limits of the measuring device.
The structure of the fundamental scientific comparative form is:

N_1 in respect X_1 equals $N_{\text{numeral 1}}$ N_{scale}
N_2 in respect X_2 equals N_{numeral} N_{scale}

The significant difference for this experiment is N_{numeral} (where either $N_1 = N_2$, or $X_1 = X_2$).

Each of the preceding examples of a comparative structure is paraphrasable in terms of this fundamental form. For example:

#097
Activity of female suspensions is 80% greater than male suspensions at pH 8.7. ↔
Female in respect of activity of suspensions at pH 8.7 equals $[?_1]_{\text{numeral}}$ $[?_2]_{\text{scale}}$.

Male in respect of activity of suspensions at pH 8.7 equals $[?_3]_{\text{numeral}}$ $[?_2]_{\text{scale}}$.

The significant difference for this experiment is [greater than the difference between the two numbers above].

$$?_1 = ?_3 + (?_3. \quad .80)$$

Though there is no statement of the exact measurement of the activity of either type of suspension, the fact that there is a sentence in which the amount of activity of one type of suspension is compared to the amount of activity of the other type of suspension presupposes that there was some determination of the amount of activity of each type of suspension independently. Therefore, the occurrence of a simple comparative sentence in the context of a scientific text implies that the measurement report sentences, which are part of the fundamental structure from which the comparative is derived, were zeroed in the formation of this text.

The measurement report sentences can actually be represented by two slightly differing structures. As more information is discovered about the structure of texts there may be structural reasons for preferring one formulation over the other. At this point, however, there are no strong criteria for making the decision between the formulations. The one we have given above (p. 65) seems slightly more intuitive. In the formulation given above the measurement sentence could vary either in the subject being measured or the respect in which it was being measured, but only one could vary in any particular case. An alternate formulation would be to specify within the structure itself which aspect of the measurement report sentences could vary. We could give the structure as:

1. N_1 in respect X_1 equals $N_{\text{numeral 1}}$ N_{scale}
   N_1 in respect X_2 equals $N_{\text{numeral 2}}$ N_{scale}

or we could specify the structure as:

2. N_1 in respect X_1 equals $N_{\text{numeral 1}}$ N_{scale}
   N^2 in respect X_2 equals $N_{\text{numeral 2}}$ N_{scale}

Either of these formulations would make it easy to state the transformations by means of which the zeroed sentences could be recovered from the comparative sentences which appear in the text. For instance, with formulation 1 we would say that the identical portions of the noun phrases in the comparative sentences become N_1, the subject being measured, and the non-identical parts of the noun phrase become X_1 and X_2, the respect in which they are being measured. A similar rule would apply if we adopted formulation 2. Though these formulations make the statement of the transformation easier, they appear counter-intuitive in some instances. Thus, in #097, given formulation 1, we get:

#097
Activity of suspensions at pH 8.7 in respect female equals $[?_1]_{\text{numeral}}$ $[?]_{\text{scale}}$.
Activity of suspensions at pH 8.7 in respect male equals $[?_2]_{\text{numeral}}$ $[?]_{\text{scale}}$.

Though this formulation of the underlying sentence follows the transformational rules, the reconstructed zeroed sentences seem intuitively wrong. This is, of course, in itself not enough reason to reject this formulation, but when we consider the overall structure of the text it seems preferable to use the more flexible formulation, which was the first one proposed. By using this more flexible formulation we may be able to make the structure in terms of which the type of text is described more uniform.

In the examples given above only the number and scale values were not explicitly recoverable from the sentence in the text. In some structures, which we will associate with this fundamental scientific comparative, not only are the number and scale values missing, but also only one of the N values is explicitly given instead of the required two. For example, a sentence which we want to call a comparative is:

#096
In the patients given homologous antigen antibodies increased.

In this instance one entire measurement report sentence and the number and scale values for the other measurement report sentence are not explicitly given in the text, but they are all reconstructable. In any scientific text if both subjects of the two fundamental measurement report sentences are not explicitly given then one sentence can be reconstructed to be:

N_1 in the untreated state (or previous state, or controls) in respect X_1 equals [$?_1$] [?].

while the other is recoverable from the text as:

N_1 in respect X_2 equals [$?_2$] [?]

That is, when a scientific text reports that a certain treatment caused a characteristic to increase, this is understood to mean that the values determined by measurement after the treatment are higher than the values would have been had the treatment not been given. There is, of course, no real method of verifying this counterfactual conditional. The concept of a control group is one way the scientist has of trying to verify such counterfactuals. It is assumed that the test animal and the control animal are identical in every respect except for the treatment which the test animal receives. Therefore the measurement of the control is evidence which can be used to discuss the effect of the treatment. In other instances, where time is not one of the assumed variables, the test animal can be measured before and after treatment and the results of the first measurement can be used as evidence for what the values of the measurement would have been had treatment not been given. It is from this point of view that *the untreated state*, *the previous state* and *controls* can be taken to be semantically equivalent possibilities for the reconstructed segments of the zeroed measurement report sentence.

The primary distinction between the comparative texts which we have labeled 1a and those which we have labeled 1b and 1c is that texts of type 1a involve only one test subject and therefore, require the reconstructed measurement report sentence with *the untreated state* as described above. Since these texts do not have two explicit groups which are being compared they do not have the standard comparative structure which requires two groups e.g.

greater than. Instead there is a high occurrence of implicit comparatives such as: increased, changed, remained constant.

Texts of type 1a often only mention the group which was treated only in the first sentence. Since it is the only group mentioned all subsequent comparatives can be understood to refer to this group. Thus a sentence like:

#259
During the infusion R.P.F. decreased.

can only be properly interpreted by looking at the first sentence of the text, in which the group being tested and the treatment given is specified. This sentence is:

The purpose of the present experiments was to observe the effect of a sustained constant infusion of Angiotensin II administered over 60 min's, on G.F.R., R.P.F. and electrolyte excretion in the anesthetized mongrel fasting dog.

Nowhere else in the text is the test group, anesthetized mongrel dog, mentioned. However, it is clear that the rest of the text can only be interpreted if we assume that each of the implicit comparatives has reference to this test animal. In the example of the implicit comparative given from this text the proper normalized form should be:

During the infusion of Angiotensin II the R.P.F. of the anesthetized mongrel fasting dog decreased.

Reconstructing the fundamental scientific comparative for this sentence would give:

Anesthetized mongrel fasting dogs in the untreated state in respect of R.P.F. equals [$?_1$] [?].
Anesthetized mongrel fasting dogs treated with Angiotensin II in respect of R.P.F. equals [$?_2$] [?].
[$?_1$] [$?_2$]

This set of sentences is a paraphrase, not simply of the original sentence, but rather of the original sentence embedded in the original text.

Once the form of the two measurement report sentences are reconstructed and the significance sentence is given, the actual form the sentence in the text takes is independent of the particular relationship between the two values which is chosen. That is, once the values in the measurement sentence are established there are very many different possible relationships between these values which can be expressed. From the point of view of the scientific text, sentences which differ only in the relationship which they express and do not differ in their reconstructed measurement report sentences or in their significance sentences will be taken to be paraphrastic. Therefore, in any semantic analysis they will be interchangeable. For example given the measurement report sentences:

Starved chickens in respect of body weight equals 3 pounds.
N-free chickens in respect of body weight equals 6 pounds.

we could derive any of the following paraphrastic sentences:

Starved chickens weigh less than *N*-free chickens.
N-free chickens weigh more than starved chickens.
The body weight of starved chickens is 50% that of *N*-free chickens.
The body weight of *N*-free chickens is twice as great as that of starved chickens.
etc.

In order to represent the comparative structure of the text, we are not interested in preserving these stylistic variations. By transforming all such sentences into their underlying scientific comparative structure we preserve the information content of this paraphrastic set while eliminating the particular relationship between the fundamental values which is reflected in the stylistic variant. In some cases the relationship between the values is expressed as movement along some scale rather than in terms of a numerical function. We find, for instance:

#197
The arterial pH rose. ↔ The arterial pH of the anesthetized mongrel fasting dog rose.

These sentences can be expressed in terms of the scientific comparative, but with the words *rise* and *fall* the zeroed measurement report sentence should read:

> N_1 at previous time in respect X_1 equals $N_{\text{numeral 1}}$ N_{scale}.
> N_1 now in respect X_1 equals $N_{\text{numeral 2}}$ N_{scale}.

These sentences, in terms of scientific comparatives, are paraphrastic with the others given above.

3.5 *Comparatives and classes of variables*

As we have mentioned before in this chapter there is a close relationship between comparative structures and measurement sentences. As we have shown above comparative structures can be considered to be the result of transformations and zeroing on two underlying measurement report sentences and a significance sentence. The occurrence of a measurement report sentence implies that a measurement was performed. So that for every comparative in a text there is an implied measurement sentence. However,given a measurement sentence there is no way to reconstruct the missing comparative. Therefore, since the comparative structure is not recoverable from the measurement sentence, any noun phrase which occurs as the subject of a measurement sentence must also occur in a comparative structure. The structure of the controlled experiment type of text demands an obligatory connection between the occurrence of a subject noun phrase in a measurement sentence and in a comparative structure. Any controlled experiment text in which one of these sentences is missing is a deficient text. It can be made an acceptable text by the inclusion of the missing sentence. In the sample texts under consideration we find, for instance, the measurement sentences:

#235
Liver samples from hens fed the basal diet were analyzed for fatty acid distribution by gas-liquid chromatography.

Heart samples from hens fed the basal diet were analyzed for fatty acid distribution by gas-liquid chromatography.

Liver samples from hens fed the basal diet supplemented with 2.5% of crude cottonseed oil were analyzed for fatty acid distribution by gas-liquid chromatography.

Heart samples from hens fed the basal diet supplemented with 2.5% of crude cottonseed oil were analyzed for fatty acid distribution by gas-liquid chromatography.

One of the comparative sentences we find is:

Liver tissues of hens fed cottonseed oil contained more stearic acid than liver tissues of hens fed the basal diet only.

We do not find any sentence of the form:

Heart tissue of hens fed cottonseed oil V Aer than heart tissue of hens fed the basal diet only.

Therefore this text is a deficient text. It can be made an acceptable text either by eliminating the two measurement sentences which contain *heart* in the noun phrase, or by including some comparative structure in which *heart* occurs in the noun phrase.

3.6 *Container structures*

The last sentence of a controlled experiment text is a container sentence. This sentence interprets the meaning of the results of the experiment which was performed. The subject of the container verb is either a pro-word (*this*) which refers to the whole previous text, or a classified word which refers to the whole previous text (*the facts, the evidence, these findings*). The structure of these container sentences is:

$$\left\{\begin{matrix}\text{This}\\ \text{These}\\ \text{The}\end{matrix}\right\}\left\{\begin{matrix}\text{evidence}\\ \text{facts}\\ \text{data}\\ \text{experiment}\end{matrix}\right\}\left\{\begin{matrix}\text{indicates}\\ \text{suggests}\\ \text{shows}\end{matrix}\right\}\text{ that } N_1\ V_1\ N_2.$$

In some of these sentences the verb phrase is missing. In these cases the zeroed verb can always be reconstructed. For instance, in the sentence:

#109
This suggested alternate metabolic pathways of hydroxylation and deamination.

There is no verb phrase. Instead this would have to be represented as:

$$N_1 \; V_{\text{cont.}} \; N_2$$

The verb phrase is restorable and we get:

This suggests that alternate metabolic pathways of hydroxylation and deamination [existed].

The sentence which occurs with the container verb is sometimes a paraphrase of some previous part of the text, and other times introduces phrases which have not occurred anywhere else in the text. An example of the first kind of sentence which is simply a restatement of the results of the experiment is:

#193
This suggests that Ca++ is important in twitch potentiation and in excitation-contraction coupling.

An example of the container structure in which new terms are encountered is:

#235
The data indicate that cottonseed oil interferes with fatty acid metabolism in the hen so that linoleic acid is converted to stearic acid rather than to oleic acid.

The occurrence of this container sentence is obligatory for texts of this type. Texts without this container sentence are incomplete and do not provide an interpretation of the results of the experiment. In order to be complete texts they must provide a sentence of this type even if it only reiterates the findings reported in the previous part of the text.

3.7 *Summary of controlled experiment structure*

Controlled experiment texts begin with a sentence which establishes the subject being used in the experiment and the treatment being performed on that subject. In some texts there is an additional sentence or two in which the technique is specified in more detail, but this sentence is optional. There is sometimes a sentence in which the measurements which were performed and the method of measurement which was used are specified. This sentence is optional, but if it does occur then every subject noun phrase in such a measurement sentence must occur in some comparative later in the text. The difference between subtype 1a and the other subtypes is that in type 1a one of the fundamental measurement report sentences in terms of which all of the comparatives in the scientific text are expressed is zeroed. It is reconstructable with the phrases: *control* or *previous state* etc. The final sentence of the text is a container sentence in which either information mentioned previously in the text is reiterated or in which some law-like statement which is supported by the evidence of the previous text is stated.

The structure which a controlled experiment text has after normalization and restoration of deleted and zeroed material is:

$$\begin{array}{l} N_1 \text{ be } Ven \\ \{N_2 \text{ be } Ven\} \end{array} \quad \text{or} \quad \begin{array}{l} N_1 \text{ be } Ven_1 \leftrightarrow (Ven)_{1\,\mathrm{adj}}\ N_1 \\ N_1 \text{ be } Ven_2 \leftrightarrow (Ven)_{2\,\mathrm{adj}}\ N_1 \end{array}$$

$$\left[\begin{array}{l}
N_1 \text{ in respect } X_1 \text{ be} \left\{\begin{array}{l}\text{determined}\\ \text{measured}\\ \text{analyzed}\\ \text{etc.}\end{array}\right\} \text{by } N_{\mathrm{method}} \\
N_2\ (N_1 \text{ in untreated state, controls, } N_1 \text{ in previous state} \\
\text{in respect } X_1 \text{ be} \left\{\begin{array}{l}\text{determined}\\ \text{measured}\\ \text{analyzed}\\ \text{etc.}\end{array}\right\} \text{by } N_{\mathrm{method}} \\
N_1 \text{ in respect } X_1 \text{ equals } N_{\mathrm{numeral}\,1}\ N_{\mathrm{scale}} \\
N_2\ (N_1 \text{ in untreated state, controls, } N_1 \text{ in previous state}) \\
\text{in respect } X_1 \text{ equals } N_{\mathrm{numeral}\,2}\ N_{\mathrm{scale}} \\
\text{The significant difference equals } N_{\mathrm{numeral}\,3}
\end{array}\right]$$

$$\left\{\begin{array}{l}\text{This}\\ \text{These}\\ \text{The}\end{array}\right\} \left\{\begin{array}{l}\text{data}\\ \text{experiment}\\ \text{evidence}\\ \text{facts}\\ \text{etc.}\end{array}\right\} V_{\text{container}} \text{ that } S$$

The five sentences enclosed by the square brackets may occur a number of times within the same text. An example of a total text after normalization and the restoration of deleted material follows so that the total structure of the text becomes structured more uniformly and therefore will be more amenable to subsequent techniques of semantic analysis.

#006

Group 1 birds were depleted of 1/3 their initial body weight by starvation.

Group 2 birds were depleted of 1/3 their initial body weight by being fed an *N*-free diet.

Group 1 birds in respect carcass composition at weight N_1 were measured by weighing.

Group 2 birds in respect carcass composition at weight N_1 were measured by weighing.

Group 1 birds in respect carcass composition equal N_2 pounds.

Group 2 birds in respect carcass composition equal N_3 pounds.

The significant difference equals N_4.

$$|N_2 - N_3| > N_4$$
$$N_2 \# N_3$$

Group 1 birds were measured in respect to response on feeding a low protein diet by method X_1.

Group 2 birds were measured in respect to response on feeding a low protein diet by method X_1.

Control birds were measured with respect to response to feeding a low protein diet by method X_1.

Group 1 birds in respect to response on feeding a low protein diet equal N_5 N_{scale}.

Group 2 birds in respect to response on feeding a low protein diet equal N_6 N_{scale}.

Control birds in respect to response on feeding a low protein diet equal N_7 N_{scale}.

The significant difference equals N_1.

$$|N_5 - N_6| > N_8, \ |N_6 - N_7| > N_8$$
$$N_6 < N_7 < N_5$$

Group 1 and Group 2 birds in respect of dietary protein intake for maximum weight gain were measured by method X_2.

Control birds in respect of dietary protein intake for maximum weight gain were measured by method X_2.

Group 1 and Group 2 birds in respect of dietary protein for maximum weight gain equal N_9 N_{scale}.

Control birds in respect of dietary protein intake for maximum weight gain equal N_{10} N_{scale}.

The significant difference equals N_{11}.

$$|N_9 - N_{10}| < N_{11}$$
$$N_9 = N_{10} + (3\text{-}5\ \%) \cdot N_{10}$$

Group 1 and group 2 birds in respect of requirements for the amino acids methionine and lysine were measured by method X_3.

Control birds in respect of requirements for the amino acids methionine and lysine were measured by method X_3.

Group 1 and group 2 birds in respect of requirements for the amino acids methionine and lysine equal N_{12} N_{scale}.

Control birds in respect of requirements for the amino acids methionine and lysine equal N_{13} N_{scale}.

The significant difference equals N_{14}.

$$|N_{12} - N_{13}| > N_{14}$$
$$N_{12} > N_{13}$$

Group 1 and group 2 birds in respect of efficiency of utilization of amino acids methionine and lysine per unit carcass nitrogen retained were measured by method X_4.

Control birds in respect of efficiency of utilization of amino acids methionine and lysine per unit carcass nitrogen retained were measured by method X_4.

Group 1 and group 2 birds in respect of efficiency of utilization of amino acids methionine and lysine per unit carcass nitrogen retained equal N_{15} N_{scale}.

Control birds in respect of efficiency of utilization of amino acids methionine and lysine equal N_{16} N_{scale}.

The significant difference equals N_{17}.

$$|N_{15} - N_{16}| > N_{17}$$
$$N_{15} > N_{16}$$
$$f(N_{13}, N_{15}) = f(N_{14}, N_{16})$$

Group 1 and group 2 birds in respect of absolute and relative requirements of arginine and tryptophan were measured by method X_5.

Control birds in respect of absolute and relative requirements of arginine and tryptophan were measured by method X_5.

Group 1 and group 2 birds in respect of absolute and relative requirements of arginine and tryptophan equal N_{18} N_{scale}.

Control birds in respect of absolute and relative requirements of arginine and tryptophan equal N_{19} N_{scale}.

The significant difference equals N_{20}.

$$|N_{18} - N_{19}| > N_{20}$$
$$N_{18} > N_{19}$$

This indicates that

A different amino acid pattern may be optimal for the repletion of protein depleted animals.

The structural description we have given above of the controlled experiment text is generative as well as descriptive. By assigning values to the variables of this structure, and performing standard binary transformations to combine the various sentences, a controlled experiment text will be generated.

4.0 HYPOTHESIS VERIFICATION

There are two different subtypes of the hypothesis verification texts. In the first type which we will discuss, a single law-like statement is made at the beginning of the text. This may be qualified by a second law-like statement which either limits the range of the first law or is a consequence of the first law. The rest of the text provides evidence which either confirms or disconfirms the hypothesis which is embodied by the law-like statement. We shall refer to this type of text as 2a.

The other type of hypothesis verification text begins by stating more than one law-like statement. These statements all give explanations for the same phenomenon. Therefore, they are mutually contradictory. The rest of the text provides evidence which can be used to decide which of the suggested hypotheses is true. We shall refer to this type of text as 2b.

In our sample texts the following text were assigned to these types:

2a	2b
#015	#052
#068	#141
#137	
#186	
#197	
#206	
#218	
#282	

As can be seen from the above lists the type of text in which a single hypothesis is confirmed or disconfirmed is much more prevalent than the type of text which decides among multiple possible hypotheses.

4.1 *Single hypothesis texts* (2a)

These texts can be represented by the structure:

$$\left[N_{\text{person}} \begin{Bmatrix} \text{finds} \\ \text{reports} \\ \text{says} \\ \text{etc.} \end{Bmatrix} \text{that } N_1 \begin{Bmatrix} \text{causes} \\ \text{induces} \\ \text{involves} \\ \text{etc.} \end{Bmatrix} N_2 \right]$$

$$\left\{ \text{This} \left\{ \begin{matrix}\text{problem}\\ \text{observation}\\ \text{etc.}\end{matrix}\right\} \text{be} \left\{\begin{matrix}\text{investigated}\\ \text{examined}\end{matrix}\right\} \text{by } N_{\text{method}} \right\}$$

$$\left\{ \text{This} \left\{ \begin{matrix}\text{data}\\ \text{study}\\ \text{etc.}\end{matrix}\right\} \left\{\begin{matrix}\text{indicates}\\ \text{proves}\\ \text{suggests}\\ \text{etc.}\end{matrix}\right\} \text{that } S_t\ (N_1,\ N_2) \right\}$$

$$\left\langle \begin{matrix}\text{comparative or descriptive}\\ \text{section of text}\end{matrix} \right\rangle$$

$$\left\{ \text{This} \left\{ \begin{matrix}\text{data}\\ \text{study}\\ \text{etc.}\end{matrix}\right\} \left\{\begin{matrix}\text{indicates}\\ \text{proves}\\ \text{suggests}\\ \text{etc.}\end{matrix}\right\} \text{that } S_t\ (N_1,\ N_2) \right\}$$

The third structure, *This data proves*, which is indicated as optional in this description is optional only in the sense that it may appear either in this position or in last position in the text. The sentence structure enclosed by square brackets may be more than one sentence long in the text. That is, the range of the container verb may be more than one sentence long. This expanded form of the container structure is not the result of syntactic considerations at the level of the sentence, but rather comes in by way of a citation transformation which we have already mentioned in connection with parenthetical structures in the discussion of controlled experiment texts (p. 59). We will discuss this transformation more fully below. This first structure may also be repeated, in which case we would have:

$$N_{\text{person 1}} \left\{ \begin{matrix}\text{finds}\\ \text{reports}\\ \text{says}\\ \text{etc.}\end{matrix}\right\} \text{that } N_2 \left\{\begin{matrix}\text{causes}\\ \text{induces}\\ \text{involves}\\ \text{etc.}\end{matrix}\right\} N_3$$

$$N_{\text{person 4}} \left\{ \begin{matrix}\text{finds}\\ \text{reports}\\ \text{says}\\ \text{etc.}\end{matrix}\right\} \text{that } N_5 \left\{\begin{matrix}\text{causes}\\ \text{involves}\\ \text{induces}\\ \text{etc.}\end{matrix}\right\} N_6$$

The section of the structure which is included in triangular brackets, in informal terms, is the evidence which is being used to support the hypothesis. The structure of this portion of the text is somewhat like the structure of the other two types of texts which are discussed in this study.

4.2 *Hypothesis statement*

In texts of type 2 the first sentence of the text is the sentence in which the law-like statement, which the text is designed to confirm or refute, is given. In texts of type 2a the source of this hypothesis is made explicit. These hypothesis structures have two parts: 1) citation-container, 2) law-like sentence. This kind of structure occurred as the first sentence of every text in our sample. In no text did it have to be reconstructed from any other structure in the text.

4.21 *Citation-container*

The two major forms in which this structure was realized in the texts were either in the form of an explicit container structure:

#015
In 1959, Schwartz and Dameshek reported that treatment of rabbits with 6-Mercaptopurine could induce a specific immune tolerance to human albumin.

or in the form of a parenthetical citation:

#282
The 1 to 11 transformation involves reduction in cobalt valence, loss of cyanide and incorporation into 11 of and aenosine moiety, derived from ATP. (Weissbach *et a,.*, J. Biol. Chem. *236*, PC 40, 1961; Peterkofsky et al., Biochem. Biophys. Res. Comms. *5*: 213, 1961).

As we have stated before, these two forms are paraphrastic. There is a transformation which has the structure:

$$N_{\mathrm{person}} \text{ says that } S \leftrightarrow S\ (N_{\mathrm{person}})$$

Therefore, we can normalize all of these texts to have the form of the container as the first sentence. It is because of this transformation that we have to extend the range of the container as it appears in the structure of the text. In some cases the parenthetical citation occurs after a number of sentences which clearly are all to be included in the citation.
For example:

#186
Ouabain causes a net loss of potassium from isolated sartorius muscle of the frog while insulin causes a net gain of K. When ouabain and insulin are combined, the muscles lose more K than in the presence of ouabain alone. (Am. J. Physiol., *200*: 1320, 1961).

In this case it will be easy to substantiate the fact that the citation does, indeed, refer to the whole set of sentences. Each of the sentences individually has the structure which the second part of these container structures must have. This kind of multiple sentence citation structure can be ambiguous, if the range of the reference is not clear. In the case of this text we will have the structure:

$[N_{\text{person}}]$ in N_{book} says that S_1. S_2.

(NOTE: for simplicity of exposition we have not included the total form of the citation which must be accounted for in the transformation. The complete form, as mentioned in the previous discussion of citations would be: N_{person}, N_{book}, N_{date}.)

In some texts a referent occurs as the subject of the container instead of a proper name, though the occurrence of the proper name is more prevalent. We find:

#206
Previous investigators working with guinea pigs have shown that the cochlear potentials decline rapidly soon after the onset of anoxia.

The referent of the phrase *previous investigators* is a proper name. We have, however, not assumed that this referent would be reconstructable, and so we have used

N_{person} rather than $N_{\text{proper name}}$.

4.22 *Law-like sentences*

In these hypothesis structures, the full sentence which occurs with the container structure discussed above, can have two forms. The first of these is an $N\ V\ V_{\mathrm{nom}}$ structure in which all the nouns which occur are generic and in which the verb is often from a small class of causal verbs. The structure of this is:

$$N_{\mathrm{generic}} \left\{ \begin{array}{l} \text{causes} \\ \text{induces} \\ \text{stimulates} \\ \text{etc.} \end{array} \right\} V_{\mathrm{nom}}$$

For example:

#186
Ouabain causes a net loss of potassium from isolated sartorius muscle of the frog.

#197
Diethyl ether stimulates respiration.

The other structure which can occur in this position is a comparative structure. We find, for instance:

#068
The vasoconstrictor potency of ephedrine in the perfused hind limb of the cat was much higher in the presence of plasma or serum.

In terms of the fundamental comparative structure this will give:

The vasoconstrictor potency of ephedrine in the perfused hind limb of the cat in respect to prescence of plasma or serum equal X_1.
The vasoconstructor potency of ephedrine in the perfused hind limb of the cat in respect to absence of plasma or serum equal X_2.

$$X_1 > X_2$$

In texts which have a comparative structure as the full sentence in the container sentence which opens the text, the law-like statement which the text addresses itself to is:

$$N_{\mathrm{generic}} \text{ causes } X_1 \left\{ \begin{array}{c} > \\ < \\ = \end{array} \right\} X_2$$

That is, if the first sentence reports that *N* in respect *X* differs from *N* in respect *Y*, then the law which will be verified in the text is that something in particular causes this difference. In the case of text #068, for instance, there is a later sentence in the text which says:

A highly potent substance which contracts the aortic strip was found... These data indicate that the active substance might be a peptide.

In some texts there may be more than one hypothesis stated at the beginning of the text. This is different from the situation in which one hypothesis is expressed in more than one sentence. An instance of more than one hypothesis is:

#015
In 1959, Schwartz and Dameshek reported that treatment of rabbits with 6-Mercaptopurine could induce a specific immune tolerance to human albumin. Recently, we have shown that an alkylating agent, cyclophosphoramide, may inhibit primary sensitization and produce specific tolerance to egg-induced systemic anaphylaxis in the guinea pig.

This text really has three hypotheses. One in the first sentence and two in the second, which can be made explicit by decomposing the binary *and* transformation and restoring the deleted material, which would routinely be done under normalization.

It is very difficult to make explicit the relationships among these hypotheses by any structural means. The connection between *tolerance* and *sensitization*, for instance, is not a relationship which is reflected at the syntactic level of analysis. In informal terms, the later hypotheses do not contradict, but rather qualify, the first hypothesis. While the pattern of connections among the noun phrases in the controlled experiment structures could be made somewhat explicit, the connections in the hypothesis verification structure are much more loose. As we will see when we look at the middle section of this type of text, in which evidence is stated, the connections between the evidence and the hypothesis is primarily at the semantic level and not at the syntactic level.

4.23 *Alternate hypotheses*

In texts of type 2b the first sentence states the alternate possible hypotheses. They are not given in a container structure in our data, but rather occur conjoined by *or*. The structure we get is:

$$V_{\mathrm{nom}} \; V \; N_{\mathrm{generic\,1}}, \; N_{\mathrm{generic\,2}}, \text{ or } N_{\mathrm{generic\,3}}.$$

This sentence is resolvable into three law-like statements. Each of these sentences has the same causal object. When we remove the *or* during the normalization procedure we get:

$$V_{\mathrm{nom}} \quad V \quad N_{\mathrm{generic\,1}}$$
$$V_{\mathrm{nom}} \quad V \quad N_{\mathrm{generic\,2}}$$
$$V_{\mathrm{nom}} \quad V \quad N_{\mathrm{generic\,3}}$$

An example of this structure from our data is:

#052
Glucose oxidation by epidermal tissue may proceed through glycolysis and the Krebs cycle, the direct oxidative pathway, or via glyoxalate oxidation. ↔

Glucose oxidation by epidermal tissue may proceed through glycolysis and the Krebs cycle.

Glucose oxidation by epidermal tissue may proceed through the direct oxidative pathway.

Glucose oxidation by epidermal tissue may proceed through glyoxalate oxidation.

These sentences are paraphrasable by:

$$N_{\mathrm{generic\,1}} \quad V_{\mathrm{causal}} \quad V_{\mathrm{nom}}$$

$$\text{Glycolysis and the Krebs cycle} \left\{ \begin{array}{l} \text{cause} \\ \text{induce} \\ \text{stimulate} \\ \text{etc.} \end{array} \right\} \text{glucose}$$

oxidation by epidermal tissue.

From this point of view, the normalized multiple hypothesis sentence is representable by the three law-like structures:

$N_{\text{generic 1}}$ V_{causal} V_{nom}
$N_{\text{generic 2}}$ V_{causal} V_{nom}
$N_{\text{generic 3}}$ V_{causal} V_{nom}

Given this structure we can see that we have three explanations, $N_{\text{generic 1}}$, $N_{\text{generic 2}}$, and $N_{\text{generic 3}}$, for the same event, V_{nom}. The data which is given in the middle part of the text supports one of these hypotheses more than the others.

4.3 *Intention sentences*

After the introductory portion of the text in which one or more hypotheses are stated there is, in some texts, a sentence in which it is made explicit that the rest of the text will deal with the concepts expressed in the first portion of the text. These sentences usually include a statement of the method of study which will be used. While the inclusion of a sentence of this type is in accord with the rest of the structure of the text, the text is not deficient if it is not included. In this sense this sentence is very much like the optional sentence in the controlled experiment text in which the method is specified, except that the hypothesis verification texts also make explicit that this method is being employed in order to study the hypothesis. Thus, we find:

#197
The present study re-examined this problem, using intact and decerebrated cats, measuring and correlating ventilation and blood acid-base changes.

#015
We have now investigated this observation in terms of the quantitative precipitin reaction using the Preer double diffusion gel agar technique.

These sentences contain a pro-word which has as its referent the hypothesis which is given in the first part of the text. It should be noted that the range of this reference does not include the container portion of the previous part of the text, but includes only the law-like statement.

4.4 *Conclusion sentences*

In hypothesis verification texts the container sentence of the form:

$$\text{This}\left\{\begin{matrix}\text{data}\\ \text{study}\\ \text{etc.}\end{matrix}\right\}\left\{\begin{matrix}\text{indicates}\\ \text{proves}\\ \text{suggests}\\ \text{etc.}\end{matrix}\right\}\text{that } S_f\ (N_1,\ N_2)$$

may occur either at the beginning of the portion of the text in which the evidence is given or it may occur at the end of the text. In either case the sentence governed by this container is a function of the opening container of the text. For instance we find:

#197
Law-like sentence:
N person reported that
Diethyl ether stimulates respiration.

Conclusion (at the end of evidence):
Therefore, it may be concluded that
In the cat diethyl ether primarily depresses respiration.

#206
Law-like sentence:
Previous investigators have shown that
The cochlear potentials decline rapidly soon after onset of anoxia.

Conclusion (at the beginning of evidence)
The present experiments indicate that
Oxygen lack is not the primary cause of this decline.

In the first example given, the relationship between the hypothesis and the conclusion is a function of the relationship between *stimulates* and *depresses*. In structural terms the two sentences can be represented as:

$$\begin{aligned}\text{Hyp.} &= N_1 \quad V_1 \quad N_2\\ \text{Con.} &= N_1 \quad V_2 \quad N_2\end{aligned}$$

They differ only in the verb phrase. Therefore, the relationship between the two sentences is the same as the difference between the two verb phrases. In this case, since *stimulates* and *depresses* are antonyms, the conclusion is a negation of the hypothesis.

In the second example the relationship is not quite so direct. It is dependent on the scientific paraphrastic transformation:

$$N_{\text{generic}\,1}\ V_1 \text{ after } N_{\text{generic}\,2}\ V_2 \leftrightarrow$$
$$N_{\text{generic}\,2}\ V_2\ V_{\text{causal}}\ N_{\text{generic}\,1}\ V_1$$

In scientific texts, if a sentence says that one event always occurs after another event, then one of the readings of this sentence is paraphrasable by a sentence which says that the second event is caused by the first. Thus, one of the readings of the hypothesis sentence in text #206 is:

Anoxia causes the cochlear potentials to decline.

Given this paraphrase of the hypothesis sentence, and given that "oxygen lack" is a synonym for anoxia we get:

$$\text{Hyp.} = N_{\text{generic}\,1}\ V_{\text{causal}}\ V_{\text{nom}\,1}$$
$$\text{Con.} = N_{\text{generic}\,1} \text{ does not } V_{\text{causal}}\ V_{\text{nom}\,1}$$

In this case, as in the first the conclusion is a negation of the hypothesis.

In both of the examples given above the relationship between the hypothesis and the conclusion sentences could be expressed in structural terms, except for the problem of word synonymy, which is resolvable at the level of weak semantics. This is, unfortunately, not always the case. In some texts the relationship between the hypothesis and the conclusion is entirely at the level of strong semantics.

For example:

#218

Hypothesis 1

Vanillic acid is a lesser acidic catabolite.

Hypothesis 2

3,4-dihydroxymandelic acid occasionally decomposes to 3,4-dihydroxybenzoic acid.

Conclusion

Vanillic acid may be an endogenous metabolite of nor-adrenaline and adrenaline.

In order to interpret the relationship of the conclusion to the hypotheses in this case it is necessary to understand the biochemical relationships involved. There is no structural or weak semantic way in which this relationship can be established.

The subject of the container verb of these conclusion sentences is always a word which has as its referent the whole previous text, e.g. *these data, these results, these studies*. In this respect it is very much like the conclusion sentence of type 1 texts. However, the position of this sentence in type 1 texts is fixed but the position of these sentences in type 2 texts is optional. Furthermore, the full sentence in the container structure in type 2 texts is a function of the full sentence in the opening container structure of the text.

4.5 *Evidence structures*

The portion of the text which was indicated by triangular brackets in the structural diagram of type 2 texts, given on page 92 is the portion which is referred to by *evidence structures*. In this case, even more than in the case of the conclusion sentences, the relationship between the original hypothesis and the sentences in this structure is primarily at the level of strong semantics. In some texts this section contains primarily comparative structures, in other texts the structures are primarily descriptive. The texts in which the evidence structures are primarily descriptive are those in which the explicit sentence which occurs in the opening container structure is a comparative. The hypothesis of these texts then becomes:

$$[N_{\text{generic}}] \text{ causes } X_1 \begin{Bmatrix} > \\ < \\ = \end{Bmatrix} X_2 \text{ (See page 82)}$$

The rest of the text then becomes a description of N_{generic}. In texts with the usual law-like sentence for the hypothesis, the evidence is in the form of a controlled experiment designed to test the hypothesis.

4.6 *Example analysis*

#197

N_{person} reported that

Diethyl ether stimulates respiration.

This problem was re-examined by the present study by measuring ventilation of intact cats.

This problem was re-examined by the present study by measuring blood acid-base [levels] of intact cats.

This problem was re-examined by the present study by measuring ventilation of decerebrated cats.

This problem was re-examined by the present study by measuring blood acid-base [levels] of decerebrated cats.

This problem was re-examined by the present study by correlating these measurements.

Intact cats were administered diethyl ether in constant concentration through a non-return system.

Decerebrated cats were administered diethyl ether in constant concentration through a non-return system.

Intact cats in respect of expiratory volume were measured with Wedge spirometer.

Decerebrated cats in respect expiratory volume were measured with Wedge spirometer.

Intact cats treated with 10-20% ether in respect of respiratory rate equal X_1.

Intact cats in untreated state in respect of expiratory volume equal X_2.

The significant difference equals X_3

$|X_1 - X_2| > X_3$

$X_1 > X_2$

Intact cats treated with 10-20% ether in respect of tidal volume equal X_4.

Intact cats in untreated state in respect of tidal volume equal X_5.

The significant difference equals X_6.

$|X_4 - X_5| = X_6$

Intact cats treated with 10-20% ether in respect arterial pH equal X_7.

Intact cats in untreated state in respect arterial pH equal X_8.

The significant difference equals X_9.

$|X_7 - X_8| > X_9$

$X_7 > X_8$

Intact cats treated with 10-20% ether in respect PCO_2 equal X_{10}.
Intact cats in untreated state in respect PCO_2 equal X_{11}.

The significant difference equals X_{12}.

$|X_{10} - X_{11}| > X_{12}$
$X_{10} < X_{11}$

Intact cats treated with 10-20% ether after the transient phase in respect of respiratory rate equal X_{13}.

$|X_1 - X_{13}| > X_3$
$X_{13} < X_1$

Intact cats treated with 10-20% ether after the transient phase in respect tidal volume equal X_{14}.

$|X_4 - X_{14}| > X_6$
$X_{14} < X_4$

Intact cats treated with 10-20% ether in respect minute volume equal X_{15}.
Intact cats treated with 10-20% ether after the transient phase in respect minute volume equal X_{16}.

The significant difference equals X_{17}.

$|X_{15} - X_{16}| > X_{17}$
$X_{15} > X_{16}$

Intact cats treated with 10-20% ether after the transient phase in respect arterial pH equal X_{18}.

$|X_7 - X_{18}| > X_1$
$X_{18} < X_7$

Intact cats treated with 10-20% ether after the transient phase in respect PCO_2 equal X_{19}.

$|X_{10} - X_{19}| > X_{12}$
$X_{19} > X_{10}$

Intact cats treated with 10-20% ether after prolonged inhalation in respect tidal volume equal X_{20}.

$|X_{14} - X_{20}| > X_6$
$X_{20} < X_{14}$

Intact cats treated with 10-20% ether after prolonged inhalation in respect respiratory rate equal X_{21}.

$$|X_{21} - X_{13}| > X_3$$
$$X_{21} > X_{13}$$

This late tachypnea is related to the development of respiratory acidosis.
Intact cats with ether treatment discontinued in respect ventilation equal X_{22}.
Intact cats treated with 10-20% ether in respect ventilation equal X_{23}.

The significant difference equals X_{24}.

$$|X_{22} - X_{23}| > X_{24}$$

(Similar results were observed in vagotimized animals except for the absence of the transient initial increase in rate and minute volume. Note: this sentence would, when completely normalized, result in a set of comparative structures identical to those on the previous pages except for having "vagotimized" substituted for "intact").

Cats treated with 2-5% ether in respect to ventilation equal X_{25}.

Cats in untreated state in respect to ventilation equal X_{26}.

$$|X_{25} - X_{26}| = X_{24}$$

Cats treated with 2-5% ether in respect arterial pH equal X_{27}

Cats in untreated state in respect to arterial pH equal X_{28}.

$$|X_{27} - X_{28}| = X_9$$

Cats treated with 2-5% ether in respect PCO_2 equal X_{29}.

Cats in untreated state in respect PCO_2 equal X_{30}.

$$|X_{29} - X_{30}| = X_{12}$$

Cats treated with 2-5% ether in respect CO_2-ventilatory response curve equal X_{31}.
Cats in untreated state in respect CO_2-ventilatory response curve equal X_{32}.

The significant difference equals X_{33}.

$$|X_{31} - X_{32}| > X_{33}$$
$$X_{31} < X_{32}$$

Therefore
It may be concluded that
In the cat diethyl ether primarily depresses respiration.

5.0 TECHNIQUE-DESCRIPTIVE

While the first two structures of texts described in this chapter make up the bulk of scientific texts in our sample, there is one other structure which does occur. In this structure some scientific procedure or phenomenon is described. In these texts there is no experimental control and very little measurement. There were only three of these type of texts in our sample: #010, #143, and #224. The structure of these texts is determined not so much by the logical and causal interrelationships among the classes involved as by the temporal sequence of events.

5.1 *Temporal structures*

5.11 *Adverbs of time*

There are different kinds of structures which determine the temporal order of events in these texts. The first of these is the standard occurrence of adverbs of time. In these texts the temporal relations which are so important to the structure of these texts may simply be made explicit. Thus, in text #224 we find:

During N_1, N_2 V_1 early.
N_2 V_3 after 24 hours.
At that time N_3 V_4 is beginning.
After N_5, N_6 V_5.

The presence of such adverbs of time clearly fixes the order of the sentences involved. However, this temporal structure need not be carried by either the order of the sentences in the text, or by these explicit markers of time sequence. We can consider that the occurrence of any sentence with an adverb of time can be paraphrased by a sentence in which a specific time is given. That is, we can consider adverbs of time to be like comparative structures which deal with some event in respect to the time at which the event took place. It follows that the occurrence of a specific adverb of time, which in turn requires a specific order among the sentences which describe the events, can be normalized by being expressed as a relationship between time of event 1, (T_1), and time of event 2, (T_2). Since many events do not take place at an instant in time, but rather over a duration of time, the time of an event can best be represented by indicating the time at which the event began, (BT), and the time at which the event ended, (ET). In the case of instantaneous events, BT = ET. Every statement of an event can be normalized to have the structure:

$$(N_1\ V_1\ +)\ \text{BT equals}\ N_{\text{numeral}\,1},\ \text{ET equals}\ N_{\text{numeral}\,2}.$$

Any temporal relationship which is expressed in the text by the order of the sentences and the occurrence of adverbs of time can be normalized by this structure and then expressed as a function of the numbers used to measure the time. In the example we outlined from #224 we would have the structure:

During the action of micrococcal nucleases on thymus DNA, the mononucleotides appear early in the reaction and reach a plateau at about 24 hours incubation.
(Micrococcal nucleases act on thymus DNA.) BT = T_1, ET = T_2.
(Mononucleotides appear.)$_2$ BT = T_3, ET = T_4.
(Mononucleotides reach a plateau.)$_1$ BT = T_5, ET = T_6.

$T_3 > T_1$
T_3 approx. $= T_1$
$T_4 > T_1$
T_4 approx. $= T_1$
$T_5 = T_1 + 24$ hours
$T_6 < T_2$

Since every temporal relationship which is expressable in terms of adverbs of time is paraphrasable by a set of sentences of the structure given above, we can substitute the above structure for every case of the adverb of time. In this way we can express all of the temporal relationships which hold between any set of events in terms of the times at which these events take place. In this way, texts which differ merely in the particular relationship in terms of which the order of the events is described, e.g. *before* vs. *after*, could be made equivalent. This is very similar to the structural normalization of comparative structures which was described in the first part of this chapter.

5.12 *Successively modified noun phrases*

In some cases the temporal order of the events in the text is carried not by overt references to relative or absolute time, but rather by the relationship among the noun phrases in the text. If a text is describing a successive procedure in which a substance is undergoing some treatment, and then the result of this treatment is then further acted upon by some other treatment, the temporal order of the event is carried by the noun phrase. This can be expressed by the structure:

N_1 be Ven_1
then
N_1 be Ven_2
then
N_1 be Ven_3

#010
Antigen solutions are mixed with melted agar at 40°-50°.
then
Antigen solutions are allowed to gel.
then
Antigen solutions are quick frozen.

The repetition of the noun phrase in these sentences is ambiguous and must therefore be amended to have a structure which is an unambiguous paraphrase of the original text. By simply repeating the noun phrase we do not indicate that the noun is being successively treated. One way to make the successive relationship explicit is to use a pro-word reference. This method of indicating the relationship between the noun phrases is used in the texts themselves:

#010
Sections of agar stained by methylene blue and examined under light reveal a mesh of agar strands. FA staining of these sections reveals antigen localized within the strands.

However, pro-words cannot be used to unambiguously state the relationships among the noun phrases. In the first place, the length of the text and the number of possible noun phrases which could be the referent makes pro-word reference ambiguous at the level of texts. In the first example we gave, pro-word reference would give us:

Antigen solutions are mixed with melted agar at 40°-50°.
These antigen solutions are allowed to gel.
These antigen solutions are quick frozen.

In the last sentence the referent of *these antigen solutions* can be either the occurrence of *antigen solutions* in the second sentence or in the first sentence. These two different referents would result in two different readings for the set of sentences. If the referent is the first sentence, then two procedures are being described, one in which antigen solutions with agar are allowed to gel and the other in which antigen solutions with agar are quick frozen. If the referent is the second sentence then only one process with three steps is being described: first mixing, then gelling, then freezing.

Moreover, even if there is no ambiguity about which noun phrase is the referent, there still may be ambiguity about the range of the referent within this noun phrase. If the noun phrase which occurs with the demonstrative, for instance, is identical with the noun phrase which is the referent then there is no ambiguity. However, if the noun with the demonstrative is identical only with a portion of the referent noun phrase then there is ambiguity. For instance, in the example with *sections of agar* we could have:

Sections of agar stained by methylene blue and examined under light reveal a mesh of agar strands. FA staining of these sections reveals antigen localized within the strands. ↔

1. Fa staining of sections of agar ...
2. FA staining of sections of agar stained by methylene blue ...

or:

3. FA staining of sections of agar stained by methylene blue and examined under light ...

In order to make both the range of the referent and the noun phrase to which it refers unambiguous, we will, instead of using referents, repeat the whole phrase. In the example given first we would get:

Antigen solutions are mixed with melted agar at 40°-50°.
Antigen solutions mixed with melted agar at 40°-50° are allowed to gel.
Antigen solutions mixed with melted agar, which are allowed to gel, are quick frozen.

One of the typical patterns in technique-descriptive texts used to indicate temporal sequence can be expressed in terms of the discussion above as:

N_1 be Ven_1
N_1 *wh* be Ven_1 be Ven_2
N_1 *wh* be Ven_1 *wh* be Ven_2 be Ven_3

.
.
.

N_1 *wh* be Ven_1 *wh* be Ven_2 . . . *wh* be Ven_k be Ven_{k+1}.

The technique-descriptive texts can be typified by not having container structures or comparative structures. On the other hand they do have temporal structures. These temporal structures come either from the normalization of adverbs of time or from successively modified noun phrases.

6.0 ABERRANT TEXTS

Two of the texts in our sample were not describable in terms of one of the three textual types specified in this chapter. These two texts are #126 and #208. In order to describe text #126 it is necessary to use both the technique-descriptive structure and the controlled experiment structure. However, the range of each of these structures is indicated by a change in paragraph. The first paragraph of the text is of type 3, while the second paragraph is of type 1a. This suggests that in longer texts more than one type of structure may have to be used to describe the total text, but that portions of the total text will each be describable in terms of one structure. Text #208 is not describable in terms of any of the structures given. Therefore from the point of view of this study it will have to be declared to be a non-scientific text. The existence of a text which is not describable in terms of the structures which have been established in this chapter guarantees that the structures which have been established by this study are not so broad that they are trivial. They in fact do distinguish among texts. Moreover, text #208 is felt to be a non-scientific text and therefore it should not be describable in terms of a scientific structure.

4

PARAPHRASTIC ANALYSIS OF TEXTS

In the previous chapters we have discussed various procedures designed to minimize the stylistic variation among texts. These normalization procedures have been of two major types. Some transformations have been used to decompose certain structures within the original texts. These have been primarily binary transformations. Decomposing one sentence into two underlying sentences makes explicit each of the relationships expressed in the underlying sentences, and, furthermore, makes each of these relationships an independent part of the text. The sets of sentences which make up the text have further been related to certain overall structures of types of texts. In describing these types of texts certain sets of sentence structures have been used to represent structures which occur at the level of the sentence in the texts, e.g. comparatives. These standardized forms eliminate some of the paraphrastic differences between texts. After these normalizations have been performed on a text it is represented as a partially ordered set of simple sentences. However, these sentences are not as simple as kernel sentences. From a linguistic point of view many texts will exhibit similar structures, and procedures of analyzing these texts can be more uniform.

Semantic analysis of texts can be done at two levels: the analysis can attempt to determine what the text is really talking about, (strong semantics), or it can try to create a new linguistic text which is semantically equivalent to the original text, (weak semantics). This chapter will address itself to approaches of this second and more modest aim. Most of the problems in information storage and retrieval can be expressed in terms of weak semantics. These problems are concerned primarily with finding an abbreviated linguistic structure which is semantically equivalent to the original text. The original text is the starting point for the discovery of this abbreviated structure. A related problem is finding the text or set of texts from a given body of texts which is the closest semantic match to a given question or information request.

1.0 POSSIBLE TECHNIQUES

The problem of weak semantic analysis is to take a given text and by applying given procedures to this text derive a text which is either a paraphrase or an included paraphrase of the original text. In addition, there must be a procedure which will distinguish trivial from non-trivial paraphrases. We shall in this chapter outline some different approaches to the paraphrastic analysis of texts. These procedures may be defined either on the normalized text which has been established in the preceeding chapters or on the text in its original form. It is the contention of this study that even the least sophisticated techniques of analysis, e.g. frequency counts, are more accurate if they are done on the normalized text rather than the original text. It should be noted that for certain purposes of matching, the original text may be represented in this abbreviated form not by a text, but rather by some variant linguistic form, as, for instance, a list of words or phrases which can be said to define the semantic content of the text.

1.1 *Extraction and recombination*

One method of generating a text which is semantically equivalent to the original text is to select portions of the original text which can then constitute the desired semantically equivalent text. The rules to produce extract paraphrases can be defined either on the original text or on the normalized text. Furthermore, these portions of text which have been selected can be either given exactly as they appear in the text from which they are extracted or they can be further recombined to form a new text. The selection rules can be based on position of the part to be selected in the text. One can, for instance, select first sentences of texts, or last sentences. Or the selection rules can be based on the structure of the sentences chosen. One can, for instance, choose all container structures. Or the selection rules can be based on some previous analysis of the text. One can determine which words other than grammatical constants occur most frequently in the text and then choose that sentence which contains the greatest number of the most frequent words.

As we have seen from the previous study of the structure of texts it is true that the first and last sentences of a text play an important role in the organization of the text, especially in hypothesis verification texts. If we use the rule to take first and last sentences we get in original form:

#006 (controlled experiment)
Birds depleted of 1/3 their initial body weight, either by starvation or by feeding a *N*-free diet, show marked differences in carcass composition at equal body weights.

This indicates that a different amino acid pattern may be optimal for repletion of protein depleted animals.

#197 (hypothesis verification)
Diethyl ether has been reported to stimulate respiration.

Therefore, it may be concluded that in the cat diethyl ether is primarily a respiratory depressant.

#010
During studies on microbiologic antigens it became desirable to extend the use of the fluorescent antibody (FA) staining technique to the identification of antigens in aqueous solution.

The technique is rapid and allows the use of non-precipitating antibodies.

Clearly the first two examples do not produce texts which are included paraphrases of the original texts. The third example is a little better, but still is not a close paraphrase of the original text. The use of the normalized text in itself only slightly improves the situation. If we use the normalized text the referent for *this* in the first example becomes clarified to "the experiment which was performed". However, we still have no way of knowing that the range of the referent is broader than the sentence given. In the second example the use of the normalized text would make explicit the fact that the two law-like sentences were from two different sources. While this would not make the extract a paraphrase of the original text it would eliminate the seeming contradiction which arises from the extract as it is given above.

It seems clear that extraction from the text, whether normalized or not, will not in itself result in adequate paraphrases. First sentences of controlled experiment texts will merely describe the experimental situation while last sentences of these texts will indicate conclusions, but not give the evidence which led to these conclusions. First sentences of hypothesis verification texts will cite the results of some other work, while last sentences, in some cases, will be functions of these cited works. First and last sentences of descriptive texts will also not be reliable paraphrases. It would seem that extraction of whole sentences from the text, without some further structuring of these sentences is inadequate for producing paraphrases.

1.2 *Structural patterning*

If we regard the texts not merely as a sequence of sentences, but rather as a sequence of sentences within the structure of a particular type of text we may be able to use the formal patterns of the structure of the text to impose certain semantic interpretations upon the sentences of the text. By using the criteria established in the previous chapters we can perhaps more closely define the attributes of different types of texts and the semantic relationships which must hold among various structures within the text. We can then use these semantic relationships in the interpretation of the text. For instance, in chapter 2 it was pointed out that the relationship between the law-like sentence in the hypothesis and the law-like sentence in the conclusion is sometimes definable in terms of weak semantics and the structure of the sentence alone, e.g.:

#197
Hyp. Diethyl ether stimulates respiration.
Con. Diethyl ether depresses respiration.

However, in other cases there is no clear relationship between the hypothesis and the conclusion, e.g. #218, which can be established on weak semantic grounds:

#218
Hyp. 1 Vanillic acid is a lesser acidic catabolite.
Hyp. 2 3,4-dihydroxymandelic acid occasionally decomposes to 3,4-dihydroxybenzoic acid.
Con. Vanillic acid may be an endrogenous metabolite of noradrenaline and adrenaline.

In these cases in which there is no obvious connection between the hypothesis and conclusion, one can use the structure of the text itself to establish such a connection. That is, given text #218, it can be established on the grounds of the structure of this text alone that there is a relationship among the two hypotheses and the conclusion. For instance, it can be concluded that noradrenaline, adrenaline, 3,4-dihydroxymandelic acid, and 3,4-dihydroxybenzoic acid are all related compounds. In some cases the relationship between two terms which can be established on the basis of the structure of the text can be quite explicit. In the structure of the middle portion of text #197 the first sentence of the controlled experiment portion of this text establishes the classes:

1. intact cats
2. decerebrated cats

The rules of the structure of the text require that these classes occur also in comparative structures. When the classes which occur with the comparatives are determined they are found to be:

1. cats with the vagi intact
2. vagotimized cats

Therefore, given the structure of the comparatives and the classes introduced by the controlled experiment statement one can establish the following semantic equivalences for this text:

intact cats = cats with the vagi intact
semantically
decerebrated cats = vagotimized cats
semantically

This semantic equivalence is valid for only this text and does not necessarily extend to the language as a whole.

In the discussion of the different types of texts which was given in the previous chapter certain of the structures which were used to define the various different kinds of texts were described in semantic, as well as syntactic terms. In that chapter these semantic devices served primarily as mnemonic devices. The structures could have been defined strictly in syntactic terms and the semantic concepts need not have entered at all. In the light of the present chapter these semantic descriptions of the structures can form a framework for the semantic interpretation of the text. The semantic labels which were given to the various structures which comprise the text can be considered to form a semantic framework in terms of which the text can be interpreted. From this point of view the actual words which occur in the text can be considered to be the variables of the semantic framework. This would mean that some of the semantic interpretation of the text would be the result of the arrangement of the syntactic structures which occur within the text itself. Such a semantic framework might have the structure:

Controlled experiment text:
We experimented with X_1, X_2, . . . X_n
in respect to Y_1, Y_2, . . . Y_k
The results of these experiments indicate that S_1

If we use this framework and use the variables found in text #006 we get:

We experimented with birds depleted of 1/3 of their initial body weight by starvation, birds depleted of 1/3 their initial body weight by feeding an *N*-free diet, and control birds *in respect to* carcass composition at weight N_1, response to feeding a low protein diet, dietary protein intake for maximum weight gain, requirements for the amino acids methionine and lysine, efficiency of utilization of amino acids methionine and lysine per unit carcass nitrogen retained, and requirements of the amino acids arginine and tryptophan. *The results of these experiments indicate* that a different amino acid pattern may be optimal for the repletion of protein depleted animals.

This text is a good included paraphrase of the original text. Once the text is in normalized form the variables which satisfy the framework can be easily retrieved. The values for $X_1, X_2, \ldots X_n$, in the structure *We experimented with...* are the values which occur at the beginning of the original text in the structure:

$$N_1 \text{ be } Ven_1 \qquad N_1 \text{ be } Ven_1 \leftrightarrow (Ven_1)_{adj}\ N_1$$

or

$$N_2 \text{ be } Ven_1 \qquad N_1 \text{ be } Ven_2 \leftrightarrow (Ven_2)_{adj}\ N_1$$

The values for $Y_1, Y_2, \ldots Y_3$, in the structure *in respect to* are the values which occur in the comparative structures in the normalized texts.

A similar semantic framework for the hypothesis verification type of text might be:

> The assertion that $S_1, S_2 \ldots S_n$
> has been investigated.
> We experimented with $X_1, X_2, \ldots X_k$
> in respect to $Y_1, Y_2, \ldots Y_j$
>
> The results of these experiments indicate that
> $f(S_1, S_2, \ldots S_n)$

Using this framework with text #197 would give the text:

The assertion that diethyl ether stimulates respiration *has been investigated.* We experimented with intact cats administered diethyl ether in constant concentration through a non-return system, and decerebrated cats administered diethyl ether in constant concentration through a non-return system *in respect to* expiratory volume, tidal volume, arterial pH, PCO_2, minute volume, respiratory rate, and CO_2 ventilatory response curve. *The results of these experiments indicate that* false (diethyl ether stimulates respiration).

The text which is produced by this procedure is, as in the case of the previous type of text, a good included paraphrase of the original text. Moreover, as in the previous case, this text is derivable directly from the normalized text.

The establishment of such semantic frameworks can serve both as a method of storing texts in a short, but regularized form and as a method eliciting and organizing texts. If texts were always representable in a short regularized form which was a paraphrase of the original form of the text then very many different kinds of matching procedures between texts could be simplified. Requests for information could be matched against these uniform texts, for instance, much more simply than they could be matched against original texts. All in all, using semantic frameworks seems to be a promising direction for the weak semantic analysis of texts.

1.3 *Frequencies*

The most common procedures now available for establishing weak semantic equivalents for texts are based primarily on frequency counts. These frequency counts always involve considerations about whether or not the word in question is a grammatical constant or not. Any frequency count on the original form of the text does not take into account the tacit occurrence of a word or phrase in a zero occurrence. That is, if the second occurrence of a word or phrase has been zeroed because of a conjunction transformation, then the frequency count of this word or phrase on the original text will not be able to take this zero occurrence into account. Yet it is clear that the semantic interpretation of the text must include a tacit understanding of this zero occurrence of the word. Moreover, the occurrence of a referent or a classifier is not reflected in any simple frequency count on the original text. Therefore, it would seem better to use normalized texts even for such elementary procedures as frequency counts.

If we use the normalized texts for determining frequencies of words in a text, and then use these frequencies to determine the structure of a paraphrase of the original text, then certain words which occur in the standardized form of some of the structures will have to receive special status. These words will function at the level of the semantics of the whole text very much like grammatical constants function at the level of the syntax of the sentence. These terms will function in determining the semantic framework of the text and their frequency will indicate the number of times a particular structure occurs, but they will not serve to differentiate one text from another at the semantic level. Therefore, the use of normalized texts for determining frequencies which could be used as a basis for establishing included paraphrastic texts would necessitate some changes in the procedures for handling the very frequent classes, such as grammatical and semantic constants. However, once such provision were made, the results of the analysis would be much more accurate than those now used. Consider, for instance, the frequencies in both the normalized and the original form of text #006:

	original	normalized
birds	2	36
depleted	5	26
weight	4	26
starve	2	13
N-free	2	13
diet	4	23
carcass	4	4
protein	4	11
controls	3	10
animals	4	0
amino acid	2	15
absolute require.	2	4
relative require.	2	4

The relative frequencies in the two types of texts follow a different pattern. Some words which occur frequently in the original text occur seldom in the normalized text, e.g. *carcass* and *animals*. Other words have very increased frequencies in the normalized text, e.g. *birds* and *amino acids*. The pattern of frequencies in the normalized text more closely matches the semantic content of the original text than does the pattern of frequencies based on the original text.

A further refinement might be made on the procedures involving the determination of frequencies. It may be that certain portions of the semantic framework of a text are more important than others in the determination of the semantic content of the text. For instance, in the hypothesis verification texts it would seem that the two container structures play a greater role in the determination of the meaning of the text than do the comparative structures. By assigning weighted frequencies to these semantically important structures the resulting frequencies might be even more reliable than those based on the normalized text alone.

2.0 CHOICE AMONG PARAPHRASES

In establishing texts which are included paraphrases of the original text it must be kept in mind that there is no single best form for the text to take. The first output of the frequency technique described in the previous section is a list of words and not a text. It was mentioned previously in this chapter that some of the included paraphrases might be in a variant form. One of these variant forms could be a list of words. Such a list of words is often used as a semantic equivalent for the original text. In order to decide among various different possible paraphrastic linguistic structures for a given text it is necessary to consider the requirements of the system in which the text is to be used. As we have seen the normalized text can be used to generate different texts which are all included paraphrases of the original text. There may be different procedures which are applicable to different uses of the text.

2.1 *Storage and retrieval*

The problems of storage and retrieval of texts involves two different regularizations and one matching procedure. In storage and retrieval systems the problem is to store large numbers of texts in as small a space as possible and then to be able to retrieve all and only texts which are pertinent to a given request. The first problem is to transform the original text into a linguistic structure which is an included paraphrase of the original text, and which is as short as possible. In chapter 1 the definition of the concept of included paraphrase points out that certain information which is in the original text is not included in the included paraphrase of the text. Clearly the problem of generating abridged texts for economy of storage and later retrieval involves deciding first of all, which information is the most important information in the text, and second, which terms are likely to be terms found in requests for information which will be regularized and then matched against this text. It is important to distinguish between two types of terms which should be excluded from the included paraphrastic form of the text. The first of these is the type of term which is unlikely to occur in request texts. For instance, it is unlikely that the word *depleted*, which occurred 26 times in the normalized text #006, would be used in a request text, while the term *amino acid* which occurred only 15 times would be very likely to occur in a request text. The inclusion of such words which would never be retrieved would create unnecessary clutter in the system. A more serious problem is the occurrence of a term which is often in request statements, but which is not crucial to the semantic interpretation of the text in question. The inclusion of this term in the included paraphrase of this text would cause this text to be retrieved when it was not relevant to the request. In general the best form for both the included paraphrase of the text and the request to take is some form of list of words.

2.2 *Summary — Abstract*

If the included paraphrase is to serve as a summary of the text in question then different considerations must be met. In the first place, the summary itself must be a text. It cannot be a variant linguistic form. The principal difference between a text which is to serve as an informative abstract and one which is to serve as an indicative abstract is in the amount of general law-like statements and the amount of specific numerical statements which are in the included paraphrase of the text. In an indicative abstract there will be few specific measurement and method statements, and many general statements. This was the type of text which was generated by the semantic framework given at the beginning of this chapter. These included paraphrases served to outline the broad semantic areas which are covered by the text.

The informative abstract of the text will include more of the measurement and numerical statements and less of the general statements. It is clear that the normalized text can serve as a source for many different kinds of semantic frameworks which will meet the interests of the particular level of information needed. By using rules which take advantage of the different structures which determine different types of texts and also take advantage of the normalized form of the text, interesting included paraphrases can be generated.

APPENDIX I

FULL TEXTS USED IN STUDY

006

Protein and amino acid requirement for repletion of protein-depleted chickens. J. D. Summers and Hans Fisher. Rutgers University, New Brunswick, N. J.

Birds depleted of 1/3 their initial body weight, either by starvation or by feeding a *N*-free diet, show marked differences in carcass composition at equal body weights. On refeeding, the *N*-free animals respond more poorly to a low protein (14%) diet than either normal controls or the starved group. The latter's response was superior to that of the controls. Maximum weight gain of depleted birds was only achieved on a higher dietary protein intake (3-5%) than necessary for controls of the same starting weight.

For the amino acids methionine and lysine the absolute requirements were increased for the depleted animals. This was offset by a greater efficiency of utilization per unit carcass nitrogen retained, so that the relative requirement (as % diet or dietary protein) remained unchanged. For arginine and tryptophan both the absolute and relative requirements were increased considerably for the depleted animals. This indicates that a different amino acid pattern may be optimal for repletion of protein depleted animals.

010

Identification of antigens in solution by the fluorescent antibody technique. J. C. Allen (intr. by L. E. Cluff.) Dept. of Med., Johns Hopkins Univ., Baltimore, Md.

During studies on microbiologic antigens it became desirable to extend the use of the fluorescent antibody (FA) staining technique to the identification of antigens in aqueous solution. Antigen solutions are mixed with melted agar at 40-50 degrees C., allowed to gel, and quick frozen. Sections are made at 32 mu in a cryostat at −15 degrees C. Studies of human albumin (HSA), human gamma globulin (HGG), and diphtheria toxin (DT) have been done. Sections of agar stained by methylene blue and examined under light reveal a mesh of agar strands. FA staining of these sections reveals antigen either localized within the strands (HSA, DT), or present as aggregates on the surface of or between the strands (HGG). HSA has been detected in concentrations as low as 0.5 mgm/ml, HGG to 0.25 mgm/ml, and DT to 0.1 L+ unit/ml. All procedures usable in FA staining of tissue-localized antigens are applicable to this technique with the possible exception of counterstaining. Agar presumably holds these antigens by a mechanical entrapment of antigen molecules within or around agar micelles. The specificity and sensitivity of the system are, as in classical FA procedures, functions of the antibody used. The technique is rapid and allows the use of non-precipitating antibodies.

015

Specific immune tolerance to ECG albumin induced in the Guinea pig by cyclophosphoramide (cytoxan). Henry Maguire, Edward Steers and Howard Maibach (intr. by W. B. Shelley) University of Pennsylvania

In 1959, Schwartz and Dameshek reported that treatment of rabbits with 6-Mercatopurine could induce a specific immune tolerance to human albumin. Recently we have shown that an alkylating agent, cyclophosphoramide, may inhibit primary sensitization and produce specific tolerance to egg-induced systemic anaphylaxis in the guinea pig. We have now investigated this observation in terms of the quantitative precipitin reaction using the Preer double diffusion gel agar technique. 1/2 kilo guinea pigs were given a course of cyclophosphoramide (10 mgm/dX10 IP) and on the 4th, 6th and 8th days received 1 mg. crystalline egg albumin IP. Drug treated animals formed no precipitins; all controls produced good titers. At 5 1/2 weeks, 16 egg-cyclophosphoramide treated animals were given a sensitizing program duplicating the initial one. This induced no measurable precipitins in 9, trace titers in 5, and high titers in 2 (anamestic response); all controls formed anti-egg precipitins in good titer. (Supported by Armed Forces Epid. Bd.).

018

The effect of dipping eggs in testosterone propionate on the bursa of fabricius of chickens in the immediate and future generation. Bruce Glick (intr. by Frederic C. McDuffie) Miss. Agr. Exp. Sta., State College, Mississippi

Fertile eggs were removed from the incubator after 3 days of incubation and dipped in 670 mg of testosterone propionate per 100 ml of ethyl alcohol (TP) and ethyl alcohol (EA). The pointed end of the egg was dipped for 5 seconds. The size of the bursa of Fabricius, a lympho-epithelial gland, and the antibody response to Salmonella pullorum antigen were markedly reduced or eliminated in chickens hatched from TP dipped eggs. An F_1 generation was produced from birds hatched from eggs dipped in TP and EA. At hatching, the mean weight of the bursa for 20 F_1 chicks from the TP and EA groups was 31 ± 7 and 43 ± 7 mg, respectively. The differences were highly significant. There were no differences between the 2 groups in body or spleen weights. The sibs of the F_1 chicks were raised to maturity and allowed to produce an F_2 generation. At hatching, the mean weight of the bursa of 55 F_2 chicks from the TP and EA groups was 38 ± 6 and 49 ± 6 mg, resp. These differences were highly significant. The eggs used to produce the F_1 and F_2 chicks were not dipped in TP or EA. The reduced size of the bursa in the F_1 and F_2 chicks may have resulted from 1) cytoplasmic transmitted changes occurring in the P_1 chicks hatched from TP eggs, or 2) by a transference of TP-like material from hen to egg. (Supported by a grant from NIH).

021

The influence of phase on the protective potency of vaccines made from Coxiella burnetii. R. A. Ormsbee (intr. by J. J. Munoz.) Rocky Mt. Lab., NIAID, Hamilton, Mont.

The properties of purified vaccines made from Coxiella burnetii in Phase I or Phase II were compared. Groups of guinea pigs were vaccinated with decimal dilutions of vaccine, challenged, and immunity estimated in terms of fever suppression. Sera from these animals were tested for agglutinins and complement-fixing (CF) antibodies after vaccination and again following challenge. Phase I preparations were shown to be superior to Phase II preparations in protective potency, although not necessarily in the elicitation of CF antibodies.

047

Fluidloss shock. Wm. C. Nungesser & H. E. Ederstrom (Introduced by E. G. Olmstead) Dept. of Physiol. & Pharm. University of North Dakota Medical School, Grand Forks, North Dakota.

Pentobarbitalized dogs were given 33% glucose IP in a volume equal to 30 ml/kg body weight. Fluid loss shock resulted in death in about 5 hours. Hematocrit rose from a control of 45 up to 66, as blood pressure fell. T-1824 and SCN space changes were compatible with initial movement of water into the peritoneal cavity and subsequent breakdown of membrane permeability regulation. Changes also reflected mass movement of water from cells into IP fluid, which increased terminally to $\pm$ 300% of injected volume. Plasma osmolality rose initially, then fell as equilibria were approached. Peritoneal fluid osmolality fell. Terminal body fluid osmolality was elevated. Early treatment with dextran (Gentran-Baxter) or saline increased survival time. Every treated dog had larger IP volume than did the untreated. Water loss seemed the cause of death with electrolyte changes secondary, possibly affecting the CNS. Dog, rat, mouse, and guinea pig showed similar maximum tolerated dose of glucose, namely 2.5-3.0 ml/kg, indicating an acute dehydration equal to initial plasma volume. (AMA Grant 358, and UND Faculty grants).

052

Oxygen consumption by preparations of adult rat epidermis. Theodore Rosett and Joseph Fogg (intr. by E. M. Irvin) Duke Univ. Med. Ctr., Durham, N.C.

Glucose oxidation by epidermal tissue may proceed through glycolysis and the Krebs cycle, the direct oxidative pathway, or via glyoxalate oxidation. Evidence has been adduced to support these various possibilities, but the major pathway has yet to be elucidated. The low order of respiration of epidermis in vitro has previously rendered such studies difficult. Direct manometric measure was obtained of oxygen consumption by rat epidermal preparations to which ADP, DPN, and GSH or 2-mercaptoethanol had been added. Dialyzed fractions of this tissue showed a 2-10 fold stimulation of respiration when pyruvate, a-ketoglutarate, glyoxalate or glucose was added. Inorganic phosphate was required for glucose stimulation of O_2 uptake. The addition of FAD, TPN, or CoA did not affect the rate of respiration. QO_2 appeared to be inversely proportional to the age of the animal.

063
Relation between stroke volume, flow patterns and diastolic filling time in conscious dogs. A. A. Charlier, D. E. L. Wilcken, A. Guz and P. Sleight (intr. by E. Brown) Cardiovasc. Res. Inst., Univ. of Calif. Sch. of Med., San Francisco.

During marked spontaneous sinus arrhythmia, we measured the effects of variations in heart rate on the stroke volume, peak flow, peak acceleration and ejection time of the right and left ventricles of conscious dogs. Coreless Kolin sine-wave electromagnetic flow meters were implanted around the roots of the aorta and pulmonary artery a week before study. A curvilinear relationship was found when diastolic filling time was plotted against stroke volume. This relationship was similar to that found by Starling between end-diastolic volume and stroke volume. Peak flow, peak acceleration and ejection time all changed in proportion to the change in stroke volume. The behavior of the right and left ventricles was similar. These findings indicate that over a wide range of diastolic filling times (0.3 to 0.8 sec.) during a steady state in a conscious dog all the characteristics of ventricular contraction which we have measured change with the length of the diastolic filling period. (Supported by USPHS grant H-6285).

068
On a vasoactive peptide (?) *in the rabbit serum* (*SVPx*). Menachem Wurzel. (intr. by J. A. Bain), Dept. Pharm., Emory Univ., Atlanta.

Morton and Tainter (1940) found that the vasoconstrictor potency of ephedrine in the perfused hind limb of the cat was much higher in the presence of plasma or serum. We reported (1961) that the isolated rabbit aortic strip suspended in Ringer is very insensitive to ephedrine, whereas the pressor potency of ephedrine is unexpectedly much higher. Therefore, we suspected and found in rabbit serum a highly potent substance which contracts the aortic strip. In subcontractile doses; it potentiates the effect of ephedrine. A protein-free active substance, 100-1000 times more potent than the starting material, was obtained from rabbit serum by dialysis and subsequent concentration. It is stable to boiling at pH 7, quite stable to prolonged boiling in strong acid, but unstable to boiling in alkali and inactivated by trypsin. These data indicate that the active substance might be a peptide, Serum Vasoactive Peptide x (SVPx). Upon introducing 2-4 ml. of blood from the carotid artery through a polythene cannula into the organ bath of an aortic strip a strong contraction results after a letency of about 1 minute. Therefore, SVPx seems to be formed at 37° in a time as short as 1-2 minutes, which is in the order of magnitude of the blood clotting time. SVPx and similar substances, including angiotensin, will be discussed. [Partly supported by Grant H-6491, USPHS).

076

The effect of a mercurial diuretic on the toxicity of acetyl strophathidin (AcS) in dogs. A. H. Ryan, Paul Noble and E. S. Colby The Chicago Medical School, Chicago, Ill.

The toxic dose of rapidly acting AcS, shown as arrhythmias in the ECG is highly reproducible in the same dog on different days, providing a suitable reference for study of the effect of other drugs on the toxicity of cardiac glycosides. The mean I. V. toxic dose of AcS for 7 dogs (37 experiments) under pentobarbital anesthesia ranged from 0.0576 + S.E. 0.002 to 0.0841 ± S.E. 0.005 mg/kg. When a diuretic dose of 0.10 ml/kg of meralluride U.S.P. was given I.V. and followed 10 min. later with divided I.V. doses of AcS (11 dogs), the mean toxic dose of AcS did not differ significantly from that of control days. The mean cardio-toxic dose of meralluride (6 dogs) indicated by widening of QRS and elevation of T wave was 0.196 ± S.E. 0.022 mg/kg (equiv. to 7.64 mg of mercury), and the mean time required for half clearance of the drug from the heart was 58.3 ± S.E. 4.2 min. Nine dogs were given cardio-toxic doses of meralluride followed in 10 to 12 min. with divided doses of AcS. The mean toxic dose of AcS did not differ significantly from that of control days, although in three experiments, the clearance of AcS required a longer time. It appears under the conditions of these experiments, that the toxicity of AcS is not increased by an I.V. dose of meralluride larger than the diuretic dose.

096

Studies on homologous and autogenous immunization in cancer patients. W. S. Blakemore, J. M. McKenna and R. Smeltzer. Harrison Dept. Surg. Res. Schs. of Medicine, U. of Pa., Phila.

Antibodies have been detected in the sera of 30% of the patients with malignancies utilizing antigens obtained by 5 aqueous-fluorocarbon extractions of HeLa (25%) and JIII (5%) (Fed. Proc. *20*: 150, 1961).

Antigens similarly extracted from HeLa or autologous tumors were inoculated into patients with advanced malignancies. Intramuscular inoculations (1 ml each) of the fifth aqueous supernatant containing 1 mg of protein/ml. and sterilized by millipore filtration were given at 3 weekly then monthly intervals. Sera were titrated by the hemagglutination technic using sheep erythrocytes coated with HeLa cell extract or extract of one of the melanosarcomas. Hemagglutination could be inhibited by adding 20/ml of antigen used for inmunization, but not be serologically in related antigens. In the patients given homologous antigen (HeLa) antibody either increased or became evident where none had been measurable prior to inmunization. Three patients receiving autologous and one receiving homologous melano-sarcoma extract showed antibody where antibody was not measurable prior to the inoculation. One showed a secondary response.

097

Sex difference in the uptake of uric acid by rat liver preparations. Jerry Leeling and Gene F. Lata. Dept. of Biochemistry, State Univ. of Iowa, Iowa City.

Disappearance of uric acid from the medium of cell suspensions at 37°C prepared from citrate of EDTA perfused rat livers, was measured colorimetrically. Although the uptake was optimal at about pH 9, a greater sex difference observed at lower pH's. Activity of female cell suspensions 80% greater than male suspensions at pH 8.7. After incubation with uric acid cells of both sexes were re-incubated with medium free of uric acid with no release of uric acid. After 50 minutes incubation with cell suspensions of either sex, 18% of the utilized substrate (90% of initial) was detected unchanged. Neither testosterone nor B-estradiol added in vitro had any effect upon observed results with cells of either sex. Uptake differences not as significant when homogenates of unperfused livers were used. (Supported by USPHS Grant A-3079).

109

Metabolism of 3-(2-*aminobutyl*-1-C^{14}) *indole acetate [etryptamine acetate] by man.* Floyd S. Egberts, Jr. and Edward C. Daniels (intr. by G. F. Cartland.) The Upjohn Company, Kalamazoo, Michigan.

Six human subjects received 30 mg. 3-(2-aminobutyl-1-C^{14}) indole acetate before and after two weeks treatment with 30 or 90 mg./day of unlabeled drug. The mean kinetic data obtained were: plasma maximum, 4.2 hr., 0.433 y/ml.; urinary excretion; 87.5%; biliary excretion, 0.5%; fecal excretion, 5%; plasma half-life, 8.2 hr.; excretion half-time, 7.8 hr. Pre-treatment with drug had no effect on rate of absorption or excretion. Fractionation of whole urine by Sephadex and paper chromatography revealed almost complete metabolism of the drug. The major metabolites were the sulfate and glucuronide conjugates of 3-(2-aminobutyl) indol-6-ol. Unmetabolized 3-(2-aminobutyl) indole and free 3-(2-aminobutyl) indol-6-ol were excreted in small amounts. Suggestive evidence was found also for deaminated metabolites, which suggested alternate metabolic pathways of hydroxylation and deamination. In several species the psychotropic effects of 3-(2-aminobutyl) indol-6-ol were much less than those of 3-(2-aminobutyl) indol acetate.

126

Role of TPNH in progesterone biosynthesis in corpora lutea.

Kenneth Savard, Alan S. Graubert and David S. Howell. Endocrine Laboratory, Dept. of Biochemistry and Medicine, University of Miami School of Medicine, Miami, Fla.

Human and bovine luteal tissues contain the following TPN-linked dehydrogenases in the following concentrations per g. of tissues; glucose-6-phosphate, 600-3000 units; 6-phosphogluconate, 100-600 units; isocitrate, 500-1500 units; and malate, 100-300 units. The method of measurement and unitage are those of Glock and McLean (1955). These enzymes exhibit characteristic TPN-specificity and requirements for Mg^{++} and (or) Mn^{++}, and are found in the 20,000 $\times$ g supernatant fraction of homogenized fresh or frozen luteal tissue. Glucose-6-P dehydrogenase is inhibited (10-40%) by estrone, estradiol-17b, and progesterone at 2-8 $\times$ 10-5 m concentrations.

When TPN (10 umoles) plus the above substrates (10 umoles) are added to incubating luteal slices, progesterone biosynthesis is significantly increased. Maximum stimulation, however, occurs with TPN and glucose-6-P suggesting that glucose-6-P dehydrogenase is the more important of these enzymes in this stimulation. (Supported in part by Grant C-4004, U.S. Public Health Service).

137

Urea denaturation of and -chymotrypsin. Charles J. Martin (intr. by R. H. McCoy) Biochem. Dept., Sch. of Med., Univ. of Pittsburgh, Pittsburgh, Pa.

In 8 M urea and 0.2 M $CaCl_2$, the rate of absorbancy change at 293 mu (293) equaled the rate of inactivation of -chymo trypsin from pH 5.5 to 9.5. (Martin, Biochim. Biophys. Acta, *51*, 610 (1961)). These changes paralleled the apparent first order constants for irreversible inactivation (determined by dilution assay) only at pH 7.0 or greater, i.e., the pH range wherein complete loss of activity occurred. Below this pH, the irreversible inactivation rate was less than the rate of reversible inactivation (determined by assay in urea). The dependence of the above effects on pH, including absorbancy changes in the region of 235 mu, has also been investigated using -chymotrypsin. The effect of urea on the two proteins was quite dissimilar. For example, the 293 for -chymotrypsin in 8 M urea and 0.2 M $CaCl_2$ was considerably faster than the rate of irreversible inactivation which, in turn, was greater than the reversible inactivation rate. The anomaly in the two inactivation rates was studied as a function of a) dilution solvent, b) temperature and pH of incubation, and c) the pH and substrate for assay. The occurrence of a substrate-induced refolding of the active site of -chymotrypsin while in urea would not be inconsistent with the results. Supported in part by USPHS Grant RG-6243.

141

The nature of catalase complex 1. G. R. Schonbaum (intr. by M. Cohn) Johnson Fndn., Univ. of Penna., Phila.

The reaction between catalase and a peroxide can be formulated as shown in equations (A), (B) or (C). A) Catalase + ROOH Complex 1 (components of peroxide retained in complex 1); BO Catalase + ROOH Complex 1 + ROH (protohemin Fe in a higher oxidation state); C) Catalase + ROOH Complex 1 + ROH. To resolve this problem the following experiments, using catalase isolated from M. lysodeikticus, were carried out with Prof. B. Chance. 1) The reversibility of the reaction between vatalase + CH_3OOH eliminated formulation (C) and a detailed study of the equilibrium was consistent with (A) but not (B). 2) Using sensitive spectrofluorometric techniques, the reaction between catalase and CH_3CH_2OOH was investigated in the presence of ADH and DPN. If equilibrium (B) obtains, the liberation of ethanol should be reflected by the reduction of DPN to DPNH; this was not observed although the method permits determination of ethanol at a level lower than 1 uM, a sensitivity sufficient to test the equilibrium of type (B). 3) Complex 1 – formed in the presence of excess of the enzyme was dissociated with KCN and the cyanide complex subsequently decomposed by addition of $AgNO_3$. In spite of a poor recovery of complex 1, this experiment and those described above support the formulation of complex 1 as a species in which components of peroxide are retained.

143

A detailed computer model of glycolysis in ascites cells. D. Garfinkel and B. Hess. (intr. by F. Karush). Johnson Fndn., Univ. of Penna., Phila., Pa.

The glycolytic system of Ehrlich ascites cells, with all enzymes of the pathway present in proper amounts and with appropriate reverse reactions, was simulated by a computer model (Univac I and II). 14 metabolites, 4 coenzymes, and 11 enzymes of glycolysis, their respective enzyme-substrate and enzyme-coenzyme complexes, and a representation of the mitochondrial oxidative phosphorylation system are included in the model, which is derived from previous work (B. Change et al., J. Biol. Chem. *235*, 2426, 1960) by adjustment of numerical values and inclusion of missing reactions. Numerical constants are determined from the experimentally observed "inhibited" steady state (with excess glucose) and then adjusted so that the model will also hold the observed endogenous steady state and follow the observed transient kinetics (B. Hess and B. Chance, J. Biol. Chem. *236*, 239, 1961; K. K. Lonberg-Holm, Biochim. Biophys. Acta *35*, 464, 1959) when glucose is added to resting cells. At least 80% of the "inhibition" in the excess-glucose steady state is due to low cytoplasmic ATP concentration (rather than product inhibition of hexokinase by glucose-6-phosphate), and this concentration also primarily determines the transient kinetics.

159

Action of glucagon on gastric HCl secretion. T. M. Lin, D. N. Benslay, W. G. Dinwiddie and G. F. Spray. Lilly Research Laboratories, Indianapolis, Indiana.

In dogs with denervated or innervated gastric pouches, glucagon significantly inhibits the secretion of HCl stimulated by meal, insulin and gastrin, but has only a slight or negligible effect on histamine &/or methacholine-stimulated secretion. The insulin-stimulated secretion is diminished by administration of glucagon &/or glucose, but glucose is less effective than glucagon for inhibition of meal-stimulated secretion when changes in blood-sugar levels caused by both are comparable. Zinc glucagon acts like glucagon on meal-stimulated secretion. A glucagon derivative devoid of blood-sugar elevating action has no effect on gastric secretion. The inhibitory effect of glucagon is not influenced by a sympathetic blocking agent, such as dibenamine or phenoxyethylamine. The mode of action of glucagon will be discussed.

186

Effect of ouabain and ouabain plus insulin on potassium efflux from frog muscle. D. R. H. Gourley Dept. of Pharmacology, Univ. of Virginia Sch. of Med., Charlottesville, Va.

Ouabain causes a net loss of potassium from isolated sartorius muscle of the frog while insulin causes a net gain of K. When ouabain and insulin are combined, the muscles lose more K than in the presence of ouabain alone (Am. J. Physiol., *200*: 1320, 1961). The efflux of K from intact sartorius muscles loaded with K_{42} has been measured in vitro under steady state conditions at 20°C. Ouabain (10^{-6} M) increases the efflux of K by 28%, which accounts for most of the net K loss caused by ouabain. A combination of ouabain (0^{-6} M) and insulin (50 mU/ml) appears to increase the K efflux somewhat less than does ouabain alone, although the difference is of borderline significance. This suggests that the major difference in the effect of ouabain and ouabain + insulin on net K movement may be in modifying the influx of K rather than the efflux.

193

Precontraction events in caffeine potentiation. M. Brust, A. Sandow and M. Sodicoff. Inst. for Muscle Disease, Inc., New York, N.Y.

Excised curarized frog sartorius muscles were massively stimulated in 1 mM caffeine-Ringer's at 24°C. This caffeine concentration raised twitch tension output to 190% of normal Ringer's values. The times from stimulus to onset of latency relaxation (L_R), to maximum depth of this relaxation (L), and to positive tension development (L_1) showed no change compared to these times in normal Ringer's. However, maximum depth of latency relaxation (R) was reduced by 11% after 15 min. in caffeine and by a further 3% during the remaining 45 min. in the test medium. Complete recovery of R occurred within 20 min. in normal Ringer's. Potentiating NO_3^--Ringer's, by contrast, shortens L_R, L and L_I and increases R, as reported elsewhere. Refractoriness did not change in caffeine, but an uncertain increase appeared in NO_3^-. Potentiating Zn^{++} and quinine, however, increase refractoriness, as shown by others. Our present results are held to mirror increased binding of Ca^{++} due to NO_3^-, and heightened Ca^{++} mobility due to caffeine, as shown by Bianchi and Shanes. The importance of Ca^{++} in twitch potentiation and in excitation-contraction coupling is thus again suggested. Supported by a grant from the Muscular Dystrophy Associations of America. Inc.

197

Effect of diethyl ether on respiration in the cat. R. L. Katz and S. H. Ngai. Depths of Anesthesiology and Pharmacology, Col. of Physicians & Surgeons, Columbia Univ., New York, N.Y.

Diethyl ether has been reported to stimulate respiration. The present study re-examined this problem, using intact and decerebrate cats, measuring and correlating ventilation and blood acid-base changes. Diethyl ether in constant concentration was administered through a non-return system. Expiratory volume was measured with a Wedge spirometer. With the vagi intact, 10-20% ether initially increased the respiratory rate with minimal change in tidal volume. The arterial pH rose and PCO_2 fell. Following this transient phase the rate, tidal volume and minute volume progressively decreased. Arterial pH fell and PCO_2 rose. Prolonged inhalation markedly depressed tidal volume. The respiratory rate increased. This late tachypnea is related to the development of respiratory acidosis. Gradual but incomplete recovery of ventilation occurred when ether was discontinued. Similar results were observed in vagotomized animals except for the absence of the transient initial increase in rate and minute volume. Two to five % ether produced minimal or no changes in ventilation, arterial pH and PCO_2, but the CO_2-ventilatory response curve was depressed. Therefore, it may be concluded that in the cat diethyl ether is primarily a respiratory depressant. (Supported by USPHS grant B31C).

206

Maintenance of cochlear potentials during anoxia. V. Honrubia, B. M. Johnstone, R. A. Butler and C. Fernandez (intr. by D. J. Ingle). Univ. of Chicago, Chicago, Ill.

Previous investigators working with guinea pigs have shown that the cochlear potentials decline rapidly soon after onset of anoxia. Indeed, if the blood supply to the cochlea is suddenly obstructed, the positive endocochlear potential is reduced to zero within the first min. Also by this time, the action potential is eliminated and the cochlear microphonic response is usually less than one-half its original value. The present experiments indicate that oxygen lack is not the primary cause of this decline. When scala vestibuli is perfused continuously during anoxia, the endocochlear potential and the cochlear microphonics can remain with little alteration for at least 10 min. and are still present after one hr. The action potential unquestionably remain longer than it does when perfusion is not carried out during anoxia. These effects seem to be largely independent of the composition of the perfusate, but they appear to be highly dependent upon the rate of perfusion. It is suggested that the role of perfusion during anoxia is to wash away some toxic agent which accumulates during anoxia and which acts upon the generators of the various cochlear potentials. (Supported by USPHS Grant B-682 (C7).)

208

Effect of technique on the shape of the salivary curve. George F. Sutherland and Ruth A. Katz, Psychophysiol. Lab., Dept. Med. Research, Spring Grove State Hosp., Balto., Md.

The salivary curve elaborated by continuous recording of parotid secretion yields the most information of any measurement of salivary flow to date. The shape of the salivary curve depends directly on the technique used to elicit salivary secretion. From a scientific standpoint the usefulness of the curve rests upon its extreme sensitivity to minute changes in the test subject and in his responses to his environment. Rigorous adherence to Pavlov's technique is required for reliable reproduction of salivary curves. Since the nervous system has no way of distinguishing between intentional and unintentional signals it responds to cues given unwittingly by the experimenter or by the apparatus itself.

Our experience leads us to conclude: (1) that any change in the subjective sense of well-being is associated with a corresponding change in the salivary curve and (2) that, conversely, when a precise technique is employed, the test may be used as a laboratory measure of changes in the subjective sense of well-being. (Supported by a grant-in-aid from the Foundation's Fund for Research in Psychiatry.)

218

The identification of vanillic acid as a catabolite of noradrenaline metabolism in the human. Lawrence Rosen, W. B. Nelson and McC. Goodall (intr. by S. R. Tipton) Memorial Research Ctr. and Hosp., Univ. of Tenn., Knoxville, Tenn.

Following infusion of *dl*-noradrenaline-2-C_{14} into 4 human subjects, the urinary catabolites were separated (Goodall et al. J. Clin. Invest. *38*, 707 (1959)). Vanillic acid (VA) has been identified among the lesser acidic catabolites. The identity of VA has been established by isotope dilution and by paper chromatography using 3 different solvent systems. Isotope dilution indicates that VA represents less than 3% of the total radioactivity recovered in the 24 hours post infusion.

The occasional decomposition of 3,4-dihydroxymandelic acid to 3,4-dihydroxybenzoic acid has been noted by Leeper *et al* (Arch. Biochem. Biophys. *77*, 417 (1958)). For this reason, attempts were made under our usual experimental procedures to determine if any portion of 3-methoxy-4-hydroxymandelic acid decomposed to VA. This decomposition was not observed. Negative results were also obtained under similar conditions with fractions containing 3-methoxy-4-hydroxyphen-ylethylglycol, free or as the sulfate. These results support the suggestion of Smith (Nature *182*, 1741 (1958)) that VA may be an endogenous metabolite of noradrenaline and adrenaline. (U.S.P.H.S. Grant H-4309).

224

Micrococcal nuclease. E. Sulkowski and M. Laskowski, Sr., Biochem. Lab. for Cancer Research, Marquette Univ. Sch. of Med., Milwaukee, Wis.

During the action of micrococcal nuclease ($15 \times 30_6$ units) on thymus DNA (150 mg), the mononucleotides, thymidylic and deoxyadenylic acids, appear early in the reaction and reach a plateau at about 24 hours incubation. At that time the liberation of the deoxycytidylic and deoxyguanylic acids is only beginning. After exhaustive digestion, 12 out of 16 dinucleotides were present but in different amounts; those starting with A and T predominated. Out of 16 trinucleotides restricted by the assumption that they start with either A or T and end with either G or C, 8 have been identified. None of the other 48 possible trinucleotides appeared. With an excess of enzyme, trinucleotides are cleaved d-XpYpZp—d-XpYp + d-Zp. Dephosphorylated trinucleotides d-XpYpZ are much more resistant and are cleaved either into d-XpYp + d-Z, or d-Xp and d-YpZ. Trinucleotides terminated in 5'-phosphate are exceedingly resistant and are cleaved d-pXpYpZ— d-pXp + d-Yp + d-Z. Hydrolysis of heptanucleotides being 3'-monophosphate by exonucleolytic action should produce 1 di- and 5 mononucleotides. Only 3 mononucleotides were found per 1 dinucleotide indicating that endonucleolytic cleavages occurred. (Supported by grants from the Atomic Energy Commission and American Cancer Society).

235

Fatty acid metabolism in hens fed crude cottonseed oil. Robert John Evans, J. A. Davidson, S. L. Bandemer, and Marjorie Anderson. Departments of Biochemistry and Poultry Science, Michigan State University, East Lansing, Mich.

Eggs produced by hens fed the basal diet plus crude cottonseed oil had larger proportions of stearic acid and smaller proportions of oleic acid than did eggs from hens fed a basal diet containing 4.1% lipide or the basal diet plus corn oil. Blood plasma, liver, heart, ovary, and depot fat samples from hens fed the basal diet or the basal diet supplemented with 2.5% of crude cottonseed oil were analyzed for fatty acid distribution by gas-liquid chromatography. Liver, plasma, and ovary of hens fed cottonseed oil contained more stearic acid and less oleic acid than similar tissues of hens fed the basal diet only. Eggs from hens fed 10% cottonseed oil contained more stearic acid than eggs from hens fed 2.5% cottonseed oil, and eggs from hens fed 2.5% cottonseed oil and 7.5% corn oil contained as much stearic acid as those from hens fed only 2.5% cottonseed oil. The data indicate that cottonseed oil interferes with fatty acid metabolism in the hen so that linoleic acid is converted to stearic acid rather than to oleic acid.

259

Effect of agiotensin-II on canine renal hemodynamics and electrolyte excretion. David P. Lauler and Roger B. Hickler (intr. by Laurence G. Wesson, Jr.) Depth.'s of Medicine, Harvard Medical School and Peter Bent Brigham Hospital, Boston, Mass.

The purpose of the present experiments was to observe the effect of a sustained constant infusion of Angiotensin-II (Ciba), administered over 60 min's, on G.F.R. (inulin), R.P.F. (P.A.H.), and electrolyte excretion in the anesthetized mongrel fasting dog. B.P. recorded via femoral artery using a Sanborn transducer. Following 3 control periods, ANG. II was infused I.V. at rates of 0.15-0.30 gamma/kg./min. causing elevations of mean arterial pressure of 25-50 mm.Hg. This infusion was continued for 4 fifteen min. periods at which time it was abruptly terminated and 3 post-infusion periods were obtained. B.P. returned to control levels 3-5 min. following termination of the infusion. Cardiac arrhythmias were occasionally observed. During the infusion G.F.R. remained constant while R.P.F. decreased causing a rise in F.F. Urine flow (V_F) was variable. In those experiments in which a rise in V_F occurred, it was associated with an increase in $U_{OSM}V$, $U_{NA}V$, $U_{Cl}V$, and an increase in negative C_{H20}. When V_F failed to rise, these parameters were decreased.

282

Conversion of vitamin B_{12} (I) to coenzyme B_{12} (II) in clostridium tetanomorphus extracts. Alan Peterkofsky and Herbert Weissbach, National Institutes of Health, Bethesda, Md.

The I to II transformation involves reduction in cobalt valence, loss of cyanide and incorporation into II of an adenosine moiety, derived from ATP (Weissbach *et al.*, J. Biol. Che. *236*, PC 40, 1961; Peterkofsky et al., Biochem. Biophys. Res. Comms. *5*: 213, 1961). ATP, GSH and yeast extract (replaceable by DPNH and FMN) (Brady and Barker, ibid *4*: 464, 1961) are required. A convenient and rapid assay for measuring formation of II from I, based on transfer of radioactivity from ATP-C_{14} to II, has been devised. The absorption decrease at 360 mu, characteristic of the I to II change, was used to observe the cobalt reduction. These studies indicate reduction of I requires ATP, GSH and yeast extract under aerobic conditions; the extent of reduction corresponds to the amount of II formed. However, anaerobically, reduction of I takes place in the absence of ATP. The formation of an easily reoxidized reduced intermediate which reacts enzymatically with ATP to form an air-stable compound appears likely. $B_{12}r$ (Diehl and Murie, Iowa State Coll. J. Sci. *26*: 555, 1952) serves as a precursor of II, but still required DPNH and FMN. Cyanide is released from C_{14} N-labelled vitamin B only when the incubation contains ATP, GSH and yeast extract. A scheme for the step-wise conversion of I to II will be presented.

APPENDIX II

PATTERNS OF PRO-WORD INTERREFERENCE IN 16 TEXTS

(NOTE: S_n is the n_{th} sentence of the text as it actually occurs.)

#006

Ref.

S_1
starved group
S_2 ↕ the latter's
S_3
S_4
S_5
⁓ ↕ this [was offset by]
S_6
S_7
S_8 this [indicates that]

#010

Ref.

S_1
S_2
S_3
S_4
S_5 sections of agar
S_6 ↕ these sections
S_7
this technique
S_8
tissue-localized antigens
↕
S_9 these antigens
S_{10}
S_{11}

#015

Ref.
S_1
S_2 ↕ We this observation
S_3 We
S_4
S_5
S_6
$\overset{\smile}{S}_7$ ↕ this

#018

Ref.
S_1
S_2
S_3
S_4
S_5
S_6
S_7
S_8
S_9
$\overset{\smile}{S}_{10}$ ↕ these differences
S_{11}

#021

Ref.
S_1
S_2 ↕ guinea pigs
S_3 these animals
S_4
S_5

#047

Ref.
S_1
S_2

S_3
S_4
S_5
S_6
S_7
S_8
S_9
S_{10}
S_{11}
S_{12}

#052

Ref.
S_1
$\breve{S}_2$ ↕ these various possibilities
S_3
S_4 ↕ rat epidermal
S_5 this tissue
S_6
S_7
S_8

#063

Ref.
S_1
S_2
S_3 ↕ a curvilinear relationship
→ S_4 this relationship
S_5
S_6
→ S_7 these findings [indicate that]

#068

Ref.

S_1

S_2 We

We

S_3 a highly potent substance

S_4 ↕ it

S_5 a protein-free active substance

S_6 ↕ it

S_7 these data indicate that

S_8

S_9

#076

Ref.

S_1

S_2

S_3

S_4

S_5

S_6

S_7 these experiments

it appears (under the conditions of) that

#097 (no pro-words)

Ref.

S_1

S_2

S_3

S_4

S_5

S_6

S_7

#109 (no pro-words)

Ref.

S_1

S_2
S_3
S_4
S_5
S_6
S_7
S_8

#126

Ref.
→S_1 glucose-6 phosphate
6-phosphogluconate
isocitrate, malate
S_2
→S_3 these enzymes
S_4
S_5
S_6
S

#137

Ref.
S_1 ↕
S_2 these changes
S_3
S_4
S_5
S^6
S_7

#143 (no pronoun reference)

Ref.
S_1
S_2
S_3
S_4

#235 (no pronoun reference)

Ref.
S_1
S_2
S_3
S_4
S_5

APPENDIX III

PATTERN OF OCCURRENCE OF WORDS (EXCEPT FOR GRAMMATICAL CONSTANTS) WHICH OCCUR TWO OR MORE TIMES IN ORIGINAL TEXTS

≠006

Birds	2	S_1			S_4				
Depleted	5	S_1			S_4	S_5		S_7	S_8
Weight	4	$2S_1$			$2S_4$				
Starve	2	S_1	S_2						
N-free	2	S_1	S_2						
Diet	4	S_1	S_2		S_4		S_6		
Carcass	2	S_1					S_6		
Protein	4		S_2		S_4		S_6		S_8
Controls	3		S_2		S^3	S^4			
Animals	4		S_2			S_5		S_7	S_8
Amino Acid	2					S_5			S_8
Absolute requirements	2					S_5		S_7	
Relative requirements	2						S_6	S_7	
		8	6	1	7	4	4	4	4

≠010

Agar	3		S_2			$2S_5$				S_9		
Antigens	6	$2S_1$	S_2				S_6		S_8	S_9		
Diptheria Toxin (DT)	3				S_4		S_6	S_7				
Fluorescent Antibody (FA)	4	S_1					S_6		S_8		S_{10}	
Stain	3	S_1					S_6		S_8			
Technique	3	S_1							S_8			S_{11}
Human Albumin (HSA)	3				S_4		S_6	S_7				
Human Gamma Globulin (HGG)	3				S_4		S_6	S_7				
Sections	3			S_3		S_5	S_6					
Strands	2					S_5	S_6					
Antibody	2										S_{10}	S_{11}
Solutions	2	S_1	S_2									
		6	3	1	3	4	9	3	4	2	2	2

$\neq$063

Right Ventricle	2	S_1					S_6	
Left Ventricle	2	S_1					S_6	
Stroke Volume	4	S_1		S_3	S_4	S_5		
Peak Flow	2	S_1				S_5		
Peak Acceleration	2	S_1				S_5		
Ejection Time	2	S_1				S_5		
Conscious Dogs	2	S_1						S_7
Relationship	2			S_3	S_4			
Diastolic Filling Time	2			S_3				$2S_7$
		7		3	2	4	2	3

$\neq$235

Eggs	5	$2S_1$			$3S_4$	
Hens	9	$2S_1$	S_2	S_3	$4S_4$	S_5
Basal Diet	6	$3S_1$	$2S_2$	S_3		
Cottonseed Oil	8	S_1	S_2	S_3	$4S_4$	S_5
Stearic Acid	5	S_1		S_3	$2S_4$	S_5
Oleic Acid	2	S_1				S_5
Corn Oil	2	S_1			S_4	
Liver	2		S_2	S_3		
Plasma	2		S_2	S_3		
Ovary	2		S_2	S_3		
Fatty Acid	2		S_2			S_5
		12	8	7	14	5

APPENDIX IV

DECOMPOSITION OF TEXTS

TEXT ≠006

Step 1. Removal of pro-words
e.g. he, she, it, they, we, the latter, the former, this, that
sentence 3 the latter — the starved group
that — the response
par. 2 sentence 2 this — para. 2 S_1
sentence 4 this — total previous text.

Step 2. Removal of S_1 CS_2 binaries with deletions restored

T—T_1 Protein requirement for repletion of protein-repleted chickens.

T_2 Amino acid requirement for repletion of protein-repleted chickens.

S_1 — S_{11} Birds depleted of 1/3 their initial body weight by starvation showed marked differences in carcass composition at equal body weights from birds depleted of 1/3 of their initial body weight by feeding a *N*-free diet.

S_2 —

S_{21} On refeeding the *N*-free animals respond more poorly to a low protein (14%) diet than normal controls.

S_{22} On refeeding the *N*-free animals respond more poorly to a low protein (14%) diet than the starved group.

S_3 The starved group's response was superior to the response of the controls.

S_4 Maximum weight gain of depleted birds was only achieved on a higher dietary protein intake (3-5%) than necessary for maximum weight of controls of the same starting weight.

S_{51} For the amino acid methionine the absolute requirements were increased for the depleted animals.

S_{52} For the amino acid lysine the absolute requirements were increased for the depleted animals.

S_6: S_{51} and S_{52} was offset by a greater efficiency of utilization per unit carcass nitrogen retained, so that the relative requirement (0.5% diet or dietary protein) remained unchanged.

S_7: S_{71} For arginine the absolute requirements were increased for the depleted animals.

S_{72} For arginine the relative requirements were increased for the depleted animals.

S_{73} For tryptophan the absolute requirements were increased for the depleted animals.

S_{74} For tryptophan the relative requirements were increased for the depleted animals.

S_5 & S_6 & S_7 indicates that a different amino acid pattern may be optimal for repletion of protein depleted animals.

NP_2 showed marked differences from NP_2

N V by $N_\varnothing$ by V_{nom}

Birds were depleted causol by starvation

N_1 Ved by V_{nom1} differed from N_1 Ved by V_{nom2}.

S_{11} Birds were depleted of 1/3 of their body weight by starvation.

S_{12} Birds were depleted of 1/3 of their body weight by being fed an N-free diet.

S_{13} The depleted starved birds were different from the depleted N-free birds.

S_{14} The difference was in carcass composition.

S_{211} The N-free animals were refed a low protein diet.

S_{212} The normal controls were fed a low protein diet.

S_{213} The N-free animals respond to the low protein diet (in S_{21}).

S_{214} The normal controls respond to the (in S_{22}) low protein diet.

S_{215} The response (in S_{23}) of the N-free animals is poorer than the response (in S_{24}) of the normal controls.

S_{221} The N-free animals were refed a low protein diet.

S_{222} The starved group were refed a low protein diet.

S_{223} The N-free animals respond to the (in S_{221}) low protein diet.

S_{224} The starved group responds to the (in S_{222}) low protein diet.

S_{225} The response (in S_{223}) of the N-free animals was poorer than the response (in S_{224}) of the starved group.

S_{31} The response (in S_{224}) of the starved group was superior to the response (in S_{214}) of the normal controls.

S_{41} Maximum weight gain of depleted birds was only achieved on a higher dietary protein intake (3-5%) than necessary for maximum weight gain of controls.

S_{411} Maximum weight gain of depleted birds was only achieved on a dietary protein intake of X.

S_{412} Maximum weight gain of controls was only achieved on dietary protein intake of Y.
S_{413} X was higher (3-5%) than Y.
S_{42} Controls starting weight is the same as depleted bird starting weight.
S_{51} The absolute requirements for the amino acid methionine were increased for the depleted animals.
S_{52} The absolute requirements for the amino acid lysine were increased for the depleted animals.

NOTE: this is not the same scientific passive as occurred in the previous paragraph. This does not go to

The experimenters increased the requirements

but rather

The experiment increased the requirements
i.e. a causal passive formal criteria for

NOTE: "increased" is an implied comparative i.e. the requirement in this situation is greater than the requirement in some other situation. Two equivalent ways to interpret the situation being compared to:

1) before depletion
2) as opposed to normal controls

The structure of the text argues for "as opposed to normal controls".

S_{511} The depleted animals require the amino acid methionine.
S_{512} The normal controls require the amino acid methionine.
S_{513} The requirement for the depleted animals is (increased over) / (greater than)
The requirement for the normal controls.
S_{521} The depleted animals require the amino acid lysine.
S_{522} The normal controls require the amino acid lysine.
The requirement (in S_{51}) for the depleted animals in
(greater than) / (increased over) the requirement (in S_{52}) for the normal controls.
S_{61} The depleted animals utilize methionine per unit carcass nitrogen retained.
S_{62} The normal controls utilize methionine per unit carcass nitrogen retained.

S_{63} The utilization of methionine (in S_{61}) for the depleted animals is of greater efficiency than the utilization of methionine (in S_{62}) for the normal controls.

S_{64} The depleted animals utilize lysine per unit carcass nitrogen retained.

S_{65} The normal controls utilize lysine per unit carcass nitrogen retained.

S_{66} The utilization of lysine (in S_{64}) for the depleted animals is of greater efficiency than the utilization of lysine (in S_{65}) for the normal controls.

S_{67} S_5 & S_6 so that:
The relative requirement for methionine and lysine of the depleted animals is unchanged from the relative requirement for methionine and lysine of the normal controls.

S_{71} The absolute requirements for arginine were increased considerably for the depleted animals.

S_{72} The relative requirements for arginine were increased considerably for the depleted animals.

S_{73} The absolute requirements for tryptophan were increased considerably for the depleted animals.

S_{74} The relative requirements for tryptophan were increased considerably for the depleted animals.

S_8 [The whole text] indicates that:

S_{81} The amino acid pattern optimal for the repletion of protein depleted animals differs from the amino acid pattern for normal controls.

TEXT ≠235

S_1 N_1 *V*en by NP_2 had NP_3 (er) and NP_4 (er) than did N_5P or N_6.

S_{11} Eggs produced by hens fed the basal diet plus crude cottonseed oil had larger proportions of stearic acid than did eggs from hens fed a basal diet containing 4.1% lipide.

S_{12} Eggs produced by hens fed the basal diet plus crude cottonseed oil had larger proportions of stearic acid than did eggs from hens fed a basal diet plus corn oil.

S_{13} Eggs produced by hens fed the basal diet plus crude cottonseed oil had smaller proportions of oleic acid than did eggs from hens fed a basal diet containing 4.1% lipide.

S_{14} Eggs produced by hens fed the basal diet plus crude cottonseed oil had smaller proportions of oleic acid than did eggs from hens fed a basal diet plus corn oil.

S_2

S_{21} Blood plasma samples from hens fed the basal diet were analyzed for fatty acid distribution by gas chromatography.

S_{22} Liver samples from hens fed the basal diet were analyzed for fatty acid distribution by gas chromatography.

S_{23} Heart ... basal diet ...

S_{24} Ovary ... basal diet ...

S_{25} Depot fat ... basal diet ...

S_{26} Blood plasma ... the basal diet supplemented with 2.5% crude cottonseed oil ...

S_{27} Liver ... the basal diet supplemented with 2.5% crude cottonseed oil ...

S_{28} Heart ... the basal diet supplemented with 2.5% crude cottonseed oil.

S_{29} Ovary ... the basal diet supplemented with 2.5% crude cottonseed oil.

S_{210} Depot fat ... the basal diet supplemented with 2.5% crude cottonseed oil.

S_3

S_{31} Liver of hens fed cottonseed oil contained more stearic acid than similar tissues of hens fed the basal diet only.

S_{32} Plasma of hens fed cottonseed oil contained more stearic acid than similar tissues of hens fed the basal diet only.

S_{33} Ovary ...

S_{34} Liver of hens fed cottonseed oil contained less oleic acid than similar tissues of hens fed the basal diet.

S_{3}S Plasma ...

S_{36} Ovary ...

S_4 —

S_{41} Eggs from hens fed 10% cottonseed oil contained more stearic acid than eggs from hens fed 2.5% cottonseed oil.

S_{42} Eggs from hens fed 2.5% cottonseed oil and 7.5% corn oil contained as much stearic acid as [those] eggs from hens fed only 2.5% cottonseed oil.

S_5 [The data] S_1, S_2, S_3 & S_4 indicate that cottonseed oil interferes with fatty acid metabolism, in the hen, so that linoleic acid is converted to stearic acid rather than to oleic acid.

TEXT ≠063

REMOVAL OF BINARIES

S_1

S_{11} During marked spontaneous sinus arrhythmia we measured the effects of variations in heart rate on the stroke volume of the right ventricle of conscious dogs.

S_{12} During marked spontaneous sinus arrhythmia we measured the effects of variations in heart rate on the stroke volume of the left ventricle of conscious dogs.

S_{13} During marked spontaneous sinus arrhythmia we measured the effects of variations in heart rate on the peak flow of the right ventricle of conscious dogs.

S_{14} During ... peak flow ... left ...

S_{15} During ... peak acceleration ... right ...

S_{16} During ... peak acceleration ... left ...

S_{17} During ... ejection time ... right ...

S_{18} During ... ejection time ... left ...

S_2 —

S_{21} Coreless Kolin sine-wave electromagnetic flow meters were implanted around the roots of the aorta a week before study.

S_{22} Coreless Kolin sine-wave electromagnetic flow meters were implanted around the roots of the pulmonary artery a week before study.

S_3 A curvilinear relationship was found when diastolic filling time was plotted against stroke volume.

S_4 [This] A curvilinear relationship was similar to that found by Starling between end-diastolic volume and stroke volume.

S_5 redundant "all" N_1, N_2 and N_3 all V.

S_{51} Peak flow changed in proportion to the change in stroke volume.

S_{52} Peak acceleration changed in proportion to the change in stroke volume.

S_{53} Ejection time changed in proportion to the change in stroke volume.

S_{54} All changed in proportion to the change in stroke volume.

S_6 The behavior of the right and left ventricles was similar.

S_7 [These findings] S_1, S_2, S_3, S_4, S_5 and S_6 indicates that over a wide range of diastolic filling times (0.3 to 0.8 sec.) during a steady state in a conscious dog all the characteristics of ventricular contraction which we have measured change with the length of the diastolic filling period.

TEXT ≠010

Decomposition of binaries and restoration of deletions which accompanied them i.e.

S_x — S_1 CS_2 — (if C = *and*) S_1, S_2

S_1 During studies on microbiologic antigens it became desirable to extend the use of the fluorescent antibody (FA) staining technique to the identification of antigens in aqueous solution.

NOTE: the decomposition of S_2 follows the transformation
N_1 be V_1, V_2 and V_3 — N_1 be V_1 and *then*
N_1 be V_2 and *then* N_1 be V_3.

This transformation occurs only with conjoined verbs. There are some cases when it seems to occur with conjoined nouns but this is only in the case when the verbs are implicit. e.g.

I like tennis, a shower, and a thick steak. I like to play tennis, take a shower and eat a thick steak. I like to play tennis and then I like to take a shower and then I like to eat a thick steak.

These sentences are all ambiguous.

S_{21} Antigen solutions are mixed with melted agar at 40°-50°C. and then.
S_{22} The antigen solutions (in S_{21}) are allowed to set.
S_{23} Antigen solutions (in S_{22}) are quick frozen.
S_3 Sections are made in a cryostat at −15°C.
S_{41} Studies of human albumin (HSA) have been done.
S_{42} Studies of human gamma globulin (HGG) have been done.
S_{43} Studies of diptheria toxin (DT) have been done.

NOTE: The *and* in S_5 is the temporal *and*

S_{51} Sections of agar are stained with methylene blue.
S_{52} Sections of agar and *then* (in S_{51}) are examined under light.
S_{53} Sections of agar (in S_{52}) reveal a mesh of agar strands.
S_{61} FA staining of [these] (in S_5) sections reveals antigens localized within the strands (HSA, DT).

or

S_{62} FA staining of [these] (in S_5) sections reveals antigen present as aggregates on the surface of the strands (HGG).

S_{63} FA staining of these sections reveal antigen present as aggregates between the strands (HGG).

S_{71} HSA has been detected in concentrations as low as 0.5 mgm/ml.

S_{72} HGG has been detected in concentrations as low as 0.25 mgm/ml.

S_{73} DT has been detected in concentrations as low as 0.1 L + unit/ml.

S_{8} All procedures usable in FA staining of tissue-localized antigens are applicable to this technique with the possible exception of counterstaining.

S_{91} Agar presumably holds [these] tissue-localized antigens by a mechanical entrapment of antigen molecules within agar micelles.

or

S_{92} Agar presumably hold [these] tissue-localized antigens by a mechanical entrapment of antigen molecules around agar micelles.

NOTE: An alternate decomposition for *S* is the inclusive *or* in which case we get

S_{91alt} Agar presumably holds [some] of (these) tissue-localized antigens by a mechanical entrapment of antigen molecules within agar micelles.

S_{92alt} Agar presumably holds [some] of (these) tissue-localized antigens by a mechanical entrapment of antigen molecules around agar micelles.

The existence of a similar ambiguous reading for the *or* of S_6 would argue on the case of the second *or* since that would make better sense of both sets of sentences.

S_{101} The specificity of the system are, as in classical FA procedures, a function of the antibody used.

S_{102} The sensitivity of the system are, as in classical FA procedures, a function of the antibody used.

S_{111} The technique is rapid.

S_{112} The technique allows the use of non-precipitating antibodies.

TEXT ≠143

Removal of binaries S_1CS_2 with deletions restored.

S_{11} The glycolytic system of Ehrlich ascites cells, with all enzymes of the pathway present in proper amounts was simulated by a computer model (Univac 1 and 11).

S_{12} The glycolytic system of Ehrlich ascites cells, with appropriate reverse reactions was simulated by a computer model (Univac 1 and 11).

NOTE: The above decomposition may, more appropriately, be regarded as an *and* with a compound-noun, as discussed in the section on the kinds of *and S*. In the example given there the structure of the sentence itself mitigated in favor of the compound-noun *and*. This case cannot be decided on syntactic grounds. In order to decide between the two possible decompositions in this case one must know the structure of the model and the influence of each of the requirements on this structure. If the final model can be regarded as the result of conjoining the model produced by the first requirement with the model produced by the second requirement then the standard *and* decomposition is correct. If, on the other hand, the model is the result of the interaction of the individual requirements then the compound-noun *and* interpretation is correct.

S_{21} 14 metabolites are included in the model which is derived from previous work by adjustment of numerical values.
S_{22} 14 metabolites are included in the model which is derived from previous work by inclusion of missing reactions.
S_{23} 4 coenzymes are included in the model which is derived from previous work by adjustment of numerical values.
S_{24} 4 coenzymes ... inclusion of missing reactions.
S_{25} 11 enzymes of glycolysis ... adjustment of numerical values.
S_{26} 11 enzymes of glycolysis ... inclusion of missing reactions.
S_{27} The enzyme-substrate of 14 metabolites ... adjustment of numerical values.
S_{28} The enzyme-substrate of 14 metabolites ... inclusion of missing reactions.
S_{29} The enzyme-substrate of 4 coenzymes ... adj. at numerical values.
S_{210} The enzyme substrate of 4 coenzymes ... inclusion of missing reactions.
S_{211} The enzyme substrate of 11 enzymes of glycolysis ... adj. of numerical values.
S_{212} The enzyme substrate of 11 enzymes of glycolysis ... inclusion of missing reactions.
S_{213} The enzyme-coenzyme complexes of 14 metabolites ... adj. of num. values.

S_{214} The enzyme-coenzyme complexes of 14 metabolites ... inclusion of missing reactions.

S_{215} The enzyme-coenzyme complexes of 4 coenzymes ... adj. of num. values.

S_{216} The enzyme-coenzyme complexes of 4 coenzymes ... inclusion of missing reactions.

S_{217} The enzyme-coenzyme complexes of 11 enzymes of gly. ... adj. of num. values.

S_{218} The enzyme-coenzyme complexes of 11 enzymes of glycolysis are included in the model, which is derived from previous work by

NOTE: an alternate decomposition for this sentence would be to first undo the *which* phrase so that we would get

NP_1 be Ven NP_2, which be Ven by$_{causal}$ NP_3 —

NP_1 be Ven NP_2, NP_2 be Ven by$_{causal}$ NP_3

NP_3 — NP_4 and NP_5

NP_1 — NP_6 and NP_7 NP_2 be Ven by$_{causal}$ NP_4,

NP_6 — NP_8, their respective NP_9 and NP_{10} NP_2 be Ven by$_{causal}$ NP_5

NP_8 — N_{11}, N_{12} and N_{13}

NP_9 — N_9 of N_{11}, N_9 of N_{12} and N_9 of N_{13}

NP_{10} — N_{10} of N_{11}, N_{10} of N_{12} and N_{10} of N_{13}

NP_7 be Ven NP_2 N_{10} of N_{12} be Ven NP_2

N_{11} be Ven NP_2 N_{10} of N_{13} be Ven NP_2

N_{12} be Ven NP_2

N_{13} be Ven NP_2

N_9 of N_{11} be Ven NP_2

N_9 of N_{12} be Ven NP_2

N_9 of N_{13} be Ven NP_2

N_{10} of N_{11} be Ven NP_{12}

This decomposition yields 12 sentences. Two from the *which* phrase and 10 from the compound noun. If we decompose the compound noun first, which is the path we established for ourselves to begin with, then each of the 10 sentences derived from decomposing the compound noun would occur with the compound *which* phrase. Then when the which phrase is decomposed we would have 20 sentences. This would be redundant since we would then have 10 occurrences of each of the two sentences of the *which* phrase.

S_{219} A representation of the mitochondrial oxidative phosphorylation system is included in the model, which is derived from previous work by adjustment of numerical values.

S_{220} A representation of the mitochondrial oxidative phosphorylation system is included in the model, which is derived from previous work by inclusion of missing reactions.

S_{31} Numerical constants are determined from the experimentally observed "inhibited" steady state (with excess glucose).
and then

S_{32} Numerical constants (from S) are adjusted so that the model will also hold the observed.

SELECTED BIBLIOGRAPHY

Bellert, Irena. "On Certain Syntactical Properties of the English Connectives *and* and *but*", *Transformations and Discourse Analysis Papers* (University of Pennsylvania, 1966).

Borko, Harold, ed. *Automated Language Processing* (New York, 1967).

Chaplin, P. G. et. al. *Safari, an On-Line Text Processing System: User's Manual.* (= *Information System Language Studies Number Fifteen*), (The MITRE Corporation), (March, 1967).

Chomsky, Noam. *Aspects of the Theory of Syntax* (Cambridge, Mass., 1965).

Chomsky, Noam. *Syntactic Structures* (The Hague, 1957).

Garvin, Paul L. ed. *Computation in Linguistics* (Bloomington, 1966).

Gopnik, Irwin, and Myrna Gopnik. "Toward a General Theory of Style: The Law of Maximal Form", *Style,* Vol. I, No. 3 (Fall, 1967).

Harris, Zellig. "Transformational Theory", *Language,* Vol. 41, No. 3 (Jul.-Sept., 1965).

—, *Mathematical Structures in Language.* To be published.

Hiz, Henry. "The role of Paraphrase in Grammar", *Monograph Series on Languages and Linguistics,* Number 17, ed. C.T.J.M. Stuart, (April, 1964).

Joshi, Aravind, and Danuta Hiz. "Transformational Decomposition: A Simple Description of an Algorithm for Transformational Analysis of English Sentences", *Proceedings of the 2nd IFIP Int. Conference on Computational Linguistics* (Grenoble, France, Aug. 1967).

Joshi, Aravind. "Informationally Equivalent Sentences", *Proceedings of the International Conference on Information Theory* (Tokyo, Japan, 1964).

Robbins, Beverly. "Relative Clause Adjuncts of a Noun", *Transformations and Discourse Analysis Projects* (University of Penna., January 1967).

Sager, Naomi. "Informational Reduction of Texts by Syntactic Analysis", Paper presented at National Institutes of Health, Seminar on Computational Linguistics, (October, 1967).

Vendler, Zeno. "Nominalizations", *Transformations and Discourse Analysis Project* (University of Penna., 1965).

Walker, D. E. *On-Line Text Processing: Introduction and Overview.* (= *Information System Language Studies Number Twelve*), (The MITRE Corporation), (March, 1967).

Zwicky, A. M. et. al. *The MITRE Syntactic Analysis Procedure for Transformational Grammars.* Reprint of Paper Presented at the Fall Joint Computer Conference, (November-December, 1965).

Bei Fragen zur Produktsicherheit wenden Sie sich bitte an:
If you have any questions regarding product safety,
please contact:

Walter de Gruyter GmbH
Genthiner Straße 13
10785 Berlin
productsafety@degruyterbrill.com